Advances in Laser Produced Plasmas Research

Advances in Laser Produced Plasmas Research

Special Issue Editors

Maricel Agop
Stefan Andrei Irimiciuc
Viorel-Puiu Paun

MDPI • Basel • Beijing • Wuhan • Barcelona • Belgrade • Manchester • Tokyo • Cluj • Tianjin

Special Issue Editors

Maricel Agop	Stefan Andrei Irimiciuc	Viorel-Puiu Paun
"Gh. Asachi" Technical University of Iasi	Institute for Laser, Plasma and Radiation Physics	University Politehnica of Bucharest
Romania	Romania	Romania

Editorial Office
MDPI
St. Alban-Anlage 66
4052 Basel, Switzerland

This is a reprint of articles from the Special Issue published online in the open access journal *Symmetry* (ISSN 2073-8994) (available at: https://www.mdpi.com/journal/symmetry/special_issues/Advances_Laser_Produced_Plasmas_Research).

For citation purposes, cite each article independently as indicated on the article page online and as indicated below:

LastName, A.A.; LastName, B.B.; LastName, C.C. Article Title. *Journal Name* **Year**, *Article Number*, Page Range.

ISBN 978-3-03936-413-8 (Pbk)
ISBN 978-3-03936-414-5 (PDF)

© 2020 by the authors. Articles in this book are Open Access and distributed under the Creative Commons Attribution (CC BY) license, which allows users to download, copy and build upon published articles, as long as the author and publisher are properly credited, which ensures maximum dissemination and a wider impact of our publications.

The book as a whole is distributed by MDPI under the terms and conditions of the Creative Commons license CC BY-NC-ND.

Contents

About the Special Issue Editors vii

Preface to "Advances in Laser Produced Plasmas Research" ix

Stefan Andrei Irimiciuc, Florin Enescu, Andrei Agop and Maricel Agop
Lorenz Type Behaviors in the Dynamics of Laser Produced Plasma
Reprinted from: *Toxins* **2019**, *11*, 1135, doi:10.3390/sym11091135 1

Maricel Agop, Ilarion Mihaila, Florin Nedeff and Stefan Andrei Irimiciuc
Charged Particle Oscillations in Transient Plasmas Generated by Nanosecond Laser Ablation
on Mg Target
Reprinted from: *Toxins* **2020**, *12*, 292, doi:10.3390/sym12020292 15

**Florin Enescu, Stefan Andrei Irimiciuc, Nicanor Cimpoesu, Horea Bedelean,
Georgiana Bulai, Silviu Gurlui and Maricel Agop**
Investigations of Laser Produced Plasmas Generated by Laser Ablation on Geomaterials.
Experimental and Theoretical Aspects
Reprinted from: *Toxins* **2019**, *11*, 1391, doi:10.3390/sym11111391 35

Maricel Agop, Nicanor Cimpoesu, Silviu Gurlui and Stefan Andrei Irimiciuc
Investigations of Transient Plasma Generated by Laser Ablation of Hydroxyapatite during the
Pulsed Laser Deposition Process
Reprinted from: *Toxins* **2020**, *12*, 132, doi:10.3390/sym12010132 57

**Nicanor Cimpoesu, Silviu Gurlui, Georgiana Bulai, Ramona Cimpoesu, Viorel-Puiu Paun,
Stefan Andrei Irimiciuc and Maricel Agop**
In-Situ Plasma Monitoring during the Pulsed Laser Deposition of $Ni_{60}Ti_{40}$ Thin Films
Reprinted from: *Toxins* **2020**, *12*, 109, doi:10.3390/sym12010109 73

About the Special Issue Editors

Maricel Agop is a full-time university professor, who teaches physics at "Gheorghe Asachi" Technical University of Iași, Physics Department. He completed his PhD at "Al. I. Cuza" University Iași, Faculty of Physics (1981–1983), and he was awarded an honorary doctorate from "Vasile Alecsandri" University of Bacau, Romania in 2013. He has been a visiting professor at Athens University in Greece and the University of Lille in France. His main scientific interest areas are fractal physics, chaos theory, plasma physics, and applications of fractal models in biology, medicine and material science. Prof. Maricel Agop has published over 200 articles in ISI journals and has published over 20 books and book chapters with renowned publishing companies. He has a h-index of 25, with over 750 citations. He is a member of the Romania Society of Physics, the European Physical Society, the Japan Institute of Metals, The New York Academy of Science and the American Association for the Advancement of Sciences.

Stefan Andrei Irimiciuc is a scientific researcher at the National Institute for Laser, Plasma and Radiation Physics in Romania. He completed his PhD at "Al. I. Cuza" University Iași, Faculty of Physics in Romania and the University of Lille in France (2014–2017). His main scientific interests are plasma physics, laser ablation, fractal physics, plasma diagnostics and material sciences. Over the course of his career, he has been awarded several prizes and distinctions for his exceptional work in laser-matter interaction and plasma physics. Stefan Irimiciuc has published over 30 papers in international journals, with a H-index of 8 and over 140 citations.

Viorel-Puiu Paun is a full-time university professor of physics at "Politehnica" University of Bucharest, Faculty of Applied Sciences, Physics Department. His serious background (education and training) consists of a Postgraduate Diploma in Biomedical Engineering, EPFL Lausanne, Switzerland (1997), Diplôme approfondi de langue française (1996), Doctor of Sciences (PhD in Physics) at Atomic Physics Institute of Bucharest (1992), Faculty of Mathematics, University of Bucharest (1985–1990), Master in Solid State Physics, Faculty of Physics, University of Bucharest (1979), Faculty of Physics, University of Bucharest (1974–1978). His main area of scientific interest is nonlinear dynamics theory, with its applications in different physico-chemical systems (nanostructures, composites) and biological systems as research expertise, including, in particular, time series, fractals analysis and diffusion process. Viorel-Puiu Paun has published more than 150 papers in national and international journals, 85 ISI journal papers, 50 communications in national and international meetings, 434 citations without self-citations, and 17 books, chapters in books and monographs. His Hirsch factor value is currently 18. He has been a visiting professor at nine prestigious universities in Europe, including the University of Cambridge, UK (2015), École Polytechnique Fédérale de Lausanne (EPFL), Switzerland (2014), University of Florence, Italy (2011), ENSICA Toulouse, France (2007), and the School of Physical Sciences, University of Kent, Canterbury, UK (2006, 2004).

Preface to "Advances in Laser Produced Plasmas Research"

The world of laser matter interaction has known great and rapid advancements in the last few years, with a considerable increase in the number of both experimental and theoretical studies. The classical paradigm used to describe the dynamics of laser produced plasmas has been challenged by new peculiar phenomena observed experimentally, like plasma particles' oscillations, plume splitting and self-structuring behavior during the expansion of the ejected particles. The use of multiple complimentary techniques has become a requirement nowadays, as different aspects can be showcased by specific experimental approaches. To balance these non-linear effects and still remain tributary to the classical theoretical, views on laser produced plasma dynamics novel theoretical models that cover the two sides of the ablation plasma (differentiability and non-differentiability) still need to be developed.

Plasma is a strongly nonlinear dynamic system, with many degrees of freedom and other symmetries which are favorable for the development of ordered structures, instabilities and transitions (from ordered to chaotic states). For instance, even the simple analysis of a transient plasma generated by laser ablation on a pure metallic target was by means of the Langmuir probe method, which showcases the presence of an oscillatory behavior at short expansion times (<1 µs), characterized by two oscillation frequencies and a classical behavior for longer evolution times. Complex space- and time-resolved analysis is often implemented to showcase the evolution of the main plasma parameters, like the electron temperature, plasma potential, or charged particle density. Complementary optical emission spectroscopy investigations, which are of a non-invasive nature, are implemented for the analysis of plasma with a more complex composition (rocks, metallic alloys or biocompatible materials). Here, the focus changes to the exited species presented in the plasma and on parameters like overall expansion velocity or temperature and individual atomic or ionic velocity and temperatures. The values of the plasma parameter can be correlated with the physical properties of the samples, achievable by implementing several surface investigation techniques, such as X-ray diffraction (XRD), EDX, AFM and optical microscopy. The correlation can also be extended for the pulsed laser deposition process, where the properties of the plasma can have an important impact on those of the thin film, and therefore clear correlations between the properties of the thin deposited film and the intrinsic properties of the laser produced plasma are required. This is still a hot topic in the laser ablation community, and here, it is tackled by generating thin films from Ni60Ti40 in various deposition conditions. In situ plasma monitoring was implemented by means of space- and time-resolved optical emission spectroscopy, and ICCD fast camera imaging. Structural and chemical analyses were performed on the thin films, using SEM, AFM, EDS, and XRD equipment. Thus, the achievement of some correlations on how the deposition parameters influence the chemical composition of the thin films was investigated.

From a theoretical perspective, innovative theoretical models were developed on the backbone of a classical Lorenz system. A mathematical representation of a differential Lorenz system was transposed into a fractal space and reduced to an integral form by operating with the Lorenz variables simultaneously on two manifolds, generating two transformation groups, one corresponding to the space coordinates transformation and another one to the scale resolution transformation. The Lorenz system was further adapted to describe the dynamics of ejected particles as a result of laser matter interaction in a fractal paradigm. The simulations were focused on the dynamics of charged particles, and showcase the presence of current oscillations, a heterogenous velocity distribution and multi-structuring at different interaction scales. Furthermore, we attempt to improve on an already established approach to simulate the dynamics of laser ablation plasmas several approaches in the fractal paradigm of motion. Theoretical models were built for the description of laser-produced plasma dynamics expressed through fractal-type equations. The calibration of such dynamics was performed through a fractal-type tunneling effect for physical systems with spontaneous symmetry breaking, or by using various mathematical operational procedures: multi structuring of the ablation plasma by means of the fractal analysis and synchronizations of the ablation plasma entities. This revealed both the self-structuring of laser-produced plasma in two structures based on its separation on different oscillation modes and the determination of some characteristics involved in the self-structuring process. The mutual conditionings between the two structures are given as joint invariant functions on the action of two isomorph groups of SL(2R) type, through the Stoler-type transformation, explicitly given through amplitude self-modulation. The blend of the empirical proof of peculiar dynamics of laser produced plasmas in both free expansion and pulsed laser deposition conditions, with diverse theoretical approaches which make them an important collection of results in low temperature plasma and laser ablation communities.

Maricel Agop, Stefan Andrei Irimiciuc, Viorel-Puiu Paun
Special Issue Editors

Article

Lorenz Type Behaviors in the Dynamics of Laser Produced Plasma

Stefan Andrei Irimiciuc [1,2,*], Florin Enescu [3], Andrei Agop [4] and Maricel Agop [5,*]

1. National Institute for Laser, Plasma and Radiation Physics, 409 Atomistilor Street, Magurele, 077125 Ilfov, Romania
2. Institute of Physics of the Czech Academy of Sciences, Na Slovance 1991/2, 118 00 Prague, Czech Republic
3. Faculty of Physics, "Alexandru Ioan Cuza" University of Iasi, 700506 Iasi, Romania
4. Material Science and Engineering Department, "Gheorghe Asachi" Technical University of Iasi Romania, 700050 Iasi, Romania
5. Department of Physics, "Gh. Asachi" Technical University of Iasi, 700050 Iasi, Romania
* Correspondence: stefan.irimiciuc@inflpr.ro (S.A.I.); m.agop@tuiasi.ro (M.A.)

Received: 23 July 2019; Accepted: 4 September 2019; Published: 6 September 2019

Abstract: An innovative theoretical model is developed on the backbone of a classical Lorenz system. A mathematical representation of a differential Lorenz system is transposed into a fractal space and reduced to an integral form. In such a conjecture, the Lorenz variables will operate simultaneously on two manifolds, generating two transformation groups, one corresponding to the space coordinates transformation and another one to the scale resolution transformation. Since these groups are isomorphs various types isometries become functional. The Lorenz system was further adapted to describe the dynamics of ejected particles as a result of laser matter interaction in a fractal paradigm. The simulations were focused on the dynamics of charged particles, and showcase the presence of current oscillations, a heterogenous velocity distribution and multi-structuring at different interaction scales. The theoretical predictions were compared with the experimental data acquired with noninvasive diagnostic techniques. The experimental data confirm the multi-structure scenario and the oscillatory behavior predicted by the mathematical model.

Keywords: Lorenz system; fractal analysis; laser produced plasmas; plasma structuring; ionic oscillations

---

1. Introduction

Laser ablation embodies a series of phenomena with a complex interconnection between them [1]. For the better understanding of this physical process there have been multiple diagnostics techniques implemented which showcased the particle removal process [2], particle dynamics after ejection [3], plasma formation and expansion [4–6], etc. Complementary, theoretical aspects of the process have been steadily showcased in the past 30 years [7–9]. The difficulty in developing theoretical models for a multi-physics process like laser ablation comes from the combination of different elements from laser physics, solid state physics and plasma physics. Thus, an appropriate theoretical approach should be able to transition different interaction scale (ns or fs for the laser beam, fs or ps for the laser matter interaction and microsecond for the plasma expansion) and still keep a connective link between them.

The development of theoretical models that can accurately describe the ablation process has been a main direction for the understanding the underline phenomena and their interdependencies. A robust model needs to contain as input variables, laser properties, the nature of the material, ejection of the material and formation of a transient plasma. This is a difficult task as the majority of the existent theoretical approaches focus on particular sequences of the laser ablation. Therefore, we found classical theoretical approach that focused on ultra-short laser interaction with dielectric material or metals [10], ns-laser ablation [11], thermal evaporation of material, particle ejection mechanisms and plasma formation or expansion. In the past decade there have been some proposals based on a fractal paradigm [12–14]. These attempts were focused on the behavior of the laser produced plasma, mainly on peculiar results like multiple structuring during expansion [6], particle oscillations [5] and temporal and spatial distribution of some plasma parameters [13]. The advantage of representing the laser ablation process in a multifractal space is given by the ease with which we can transition between different interaction scales and different sequences of the process. This was shown for a laser produced plasma on a single element [13] and multi-component target [12]. The core premise of the model is a hydrodynamic description of the plasma plume expansion, without an explicit involvement of the experimental parameters.

In this paper we develop a mathematical model starting from classic Lorenz system and we transpose it in nondifferential (fractal) representation. From such a perspective, Lorenz type variable will depend both on space coordinates and scale resolutions. Consequently, every variable will act as a limit of a family of functions which are non-differentiable for null scale resolutions and differentiable for non-null scale resolutions; every variable will operate simultaneously on two manifolds, one generated by the space coordinates' transformation group and another one generated by the scale resolution transformation group; these two groups are isomorphs, so that various types of isometries can be applied (embeddings, compactizations, etc.). The system will be solved in a fractal space. The final aim of this development is the understanding the dynamics of a laser produced plasma with a multi-component structure, generated on a complex target. The solution allows the simulation of particle velocity distribution across a wide range of scale resolutions; the charge particle spatial distribution, showcasing the formation of a space charge double layer at the interface between plasma structure; and charged current temporal trances at various scale resolutions. The theoretical predictions are compared with experimental data extracted from a laser produced plasma on a chalcopyrite target by means of Intensified Coupled Charged Device (ICCD) fast camera imaging and optical emission spectroscopy. The theoretical simulations are confirmed by the experimental data.

2. Mathematical Model

2.1. Route to Non-Differentiability

Non-linearity and chaoticity are fundamental attributes of a Lorenz type system. Usually, the classical models built to describe the Lorenz type dynamics, are developed on the presumption of differentiability and integrability at all scale resolutions. The successes of these approaches need to be understood sequentially, on domains in which the differentiability and integrability are still respected. The differential and integral mathematical procedures fail when we attempt to describe nondifferentiable dynamics of Lorenz type system.

In order to describe the non-differentiable dynamics of a Lorenz type system and attempt to remain tributary to the well-established differentiable and integral mathematical procedures, it is necessary to introduce the scale resolution both in the variable and differentiable equations describing such dynamics. All the variables are describing a Lorenz type system from a non-differentiable perspective, which depends on both space-time coordinates and the scale resolution. Consequently, instead of operating, for example, with a single variable described by a nondifferentiable function, we will operate only with the approximations of this function obtained by mediating it at various resolution scales. This special function will act as a limit of a family of functions which are non-differentiable for zero scale resolution and differentiable for non-zero scale resolutions.

This way of describing nondifferentiable dynamics of a Lorenz type system implies both the generation of a new geometry of movement and a new class of Lorenz type models (which we will refer to as fractal/multifractal Lorenz systems). According to these geometric studies for which the movement laws, invariants of the space time coordinates' transformation are integrated over the scale resolutions which are invariant of the scale resolution's transformation. If we further admit that these geometric structures are based on the concept of multifractality, then the holographic implementation of movement—seen through scale relativity theory with an arbitrary fractal dimension or through operational procedures in the description of physical systems [14]-becomes essential when describing the dynamics of non-differentiable Lorenz type systems. However, both ways presented above imply the definition of nondifferentiable dynamics of a Lorenz type system on continuous but nondifferentiable curves. Therefore:

(i) Any variable used to describe the dynamics of a nondifferential Lorenz type system will be described through multifractal mathematical functions dependent on both the spatial and temporal coordinates, and on the scale resolution.

(ii) The laws describing these dynamics are invariant with respect to the spatial coordinates and temporal transformation, and the scale resolution transformation.

(iii) The constraints on the Lorenz type system dynamics, described through continuous and differentiable curves of a Euclidian space, are replaced by the dynamics of a system lacking any constraints, and being described by continuous and nondifferentiable curves in a multifractal space.

(iv) Between any two points in the multifractal space there is an infinity of curves describing the dynamics of a systems (its geodesics). The indiscernibility between these curves is a natural property of multifractalization through stochasticization; meanwhile, their discernibility is the result of a selection process based on the principle of maximum informational energy [14]. From such a perspective, any Lorenz type system with dynamics described by continuous and differentiable curves has hidden dissipative information (lacks memory). Otherwise, Lorenz type systems described by continuous and nondifferentiable curves have explicit information (presents memory).

Scale Resolutions

Let us consider a multifractal function (representing any of the variables describing the dynamics of a Lorenz type system) $f(u)$, defined in the closed interval $u \in [a, b]$. Let us also consider the set of values for the u variable:

$$u_a = u_0, \; u_1 = u_0 + \mu, \ldots, \; u_n + n\mu = u_b. \tag{1}$$

We will define $f(u, \mu)$ as the broken line connecting the points:

$$f(u_0), \; f(u_1), \ldots, \; f(u_n). \tag{2}$$

Then $f(u, \mu)$ becomes a μ-scale approximation.

Let us now consider $f(u, \overline{\mu})$, the $\overline{\mu}$-scale approximation of the same multifractal function. Since $f(u)$ is self-similar virtually everywhere, if $\overline{\mu}$ and μ are considered small, the approximations $f(u, \overline{\mu})$ and $f(u, \mu)$ lead to the same result when multifractal phenomena are investigated. Comparing these two situations, there is an infinitesimal increase/decrease $d\mu$ of μ that would correspond to an increase/decrease $d\overline{\mu}$ of $\overline{\mu}$, only in the case of scale contraction or dilatation. In that case, the following relationship is satisfied:

$$\frac{d\mu}{\mu} = \frac{d\overline{\mu}}{\overline{\mu}} = d\rho. \tag{3}$$

Therefore, the rational for the scale $\mu + d\mu$ and $d\mu$ needs to be constant. In these conditions we can consider the infinitesimal transformation of the scale as:

$$\mu' = \mu + d\mu = \mu + \mu d\rho. \tag{4}$$

by performing such a transformation, in the case of $f(u, \mu)$, it results in:

$$f(u, \mu') = f(u, \mu + \mu d\rho). \tag{5}$$

Furthermore, if we use stop at the first approximation of the function we get:

$$f(u,\mu') = f(u,\mu) + \frac{\partial f}{\partial \mu}(\mu' - \mu). \qquad (6)$$

Meaning,

$$f(u,\mu') = f(u,\mu) + \frac{\partial f}{\partial \mu}\mu d\rho. \qquad (7)$$

Moreover, we note that, for a fixed arbitrary μ_0,

$$\frac{\partial \ln\frac{\mu}{\mu_0}}{\partial \mu} = \frac{\partial (\ln\mu - \ln\mu_0)}{\partial \mu} = \frac{1}{\mu}, \qquad (8)$$

so that Equation (7) can be rewritten as:

$$f(u,\mu') = f(u,\mu) + \frac{\partial f(u,\mu)}{\partial \ln\left(\frac{\mu}{\mu_0}\right)} d\rho. \qquad (9)$$

In the end we will obtain:

$$f(u,\mu') = \left[1 + \frac{\partial}{\partial \ln\left(\frac{\mu}{\mu_0}\right)} d\rho\right] f(u,\mu). \qquad (10)$$

which showcases the dilatation/contraction operator:

$$\hat{O} = \frac{\partial}{\partial \ln\left(\frac{\mu}{\mu_0}\right)}. \qquad (11)$$

Equation (11) also showcases the fact that the intrinsic variation of the scale resolution is not in μ but on $\ln\left(\frac{\mu}{\mu_0}\right)$.

2.2. Non-Differentiable Lorenz Type Systems

Let us now consider the classic Lorenz system [15,16] described in a non-dimensional coordinate system by the differentiable equations:

$$\begin{aligned}\frac{d\overline{X}}{d\overline{T}} &= \sigma(\overline{X} - \overline{Y}). \\ \frac{d\overline{Y}}{d\overline{T}} &= -r\overline{X} - \overline{Y} - \overline{XZ}. \\ \frac{d\overline{Z}}{d\overline{T}} &= -b\overline{Z} + \overline{YX}.\end{aligned} \qquad (12)$$

In Equation (12) the variables are obviously differentiable and can be integrated. According to the paradigm presented above, when we consider the non-differentiable and non-integrable variables, the system of Equation (12) becomes a multifractal Lorenz type system. For this to occur, the variable needs to be dependent both on the spatial and temporal coordinates, and on the scale resolution. Moreover, the affine parameter $\overline{T}$ which characterizes the trajectories in the space phase ($\overline{X}, \overline{Y}$ and $\overline{Z}$) needs to be dependent on the resolution scale. Using such a hypothesis, the variables $\overline{X}, \overline{Y}$ and $\overline{Z}$ will operate simultaneously on two manifolds, one generated by the space-time coordinates' transformation group and another one generated by the scale resolution transformation group. Since these two group are isomorphs, or in an extremely restrictive case self-morphs, embedding type isometries, self-embedding isometries, compactization type isometries, etc., become functional.

A possible embedding scenario can be performed through scaling:

$$\overline{X} \to \frac{X}{\varepsilon}, \ \overline{Y} = \frac{Y}{\sigma\varepsilon^2}, \ \overline{Z} = \frac{Z}{\sigma\varepsilon^2}, \ \overline{T} = \varepsilon T. \tag{13}$$

with,

$$\varepsilon = \ln\frac{\mu}{\mu_0}. \tag{14}$$

In Equations (13) and (14), ε can functionally be dependent on only one fractal dimension, D_F, following the relationship $\varepsilon = \varepsilon(D_F)$. In that case, using Equation (12) we will describe the dynamics of a mono-fractal Lorenz type system. We can find another case of ε which can be dependent on a singularity spectrum f_α following the relationship $\varepsilon = \varepsilon(f_\alpha)$. In that case, Equation (12) describes the dynamics of a multifractal Lorenz type system.

The scaling Equation (13) specifies the fact that, simultaneously with the contraction in the space phase attributed to a mono-fractal or multifractal dynamics of a Lorenz type system, the dilation of time takes place. As such, the classical Lorenz system (12) through (13) and (14) becomes a multifractal Lorenz type system described by the following system:

$$\frac{dX}{dT} = Y - \varepsilon\sigma X.$$

$$\frac{dY}{dT} = X - \varepsilon Y - XZ. \tag{15}$$

$$\frac{dZ}{dT} = XY - \varepsilon b Z.$$

either through $\varepsilon = \varepsilon(D_F)$ or $\varepsilon = \varepsilon(f_\alpha)$.

In this system the dynamics variables X, Y and Z are multi fractal functions, as they depend not only on the space-time coordinates but also on the scale resolution. Therefore, instead of working with X, Y and Z variables, which are described by non-differentiable functions, we will operate only with approximations of these functions obtained by their averaging at various scale resolutions given by Equation (14). Any of the variables X, Y or Z will behave like limits of a family of functions, which are non-differentiable on null scale resolutions ($\varepsilon \to 0$, and $\mu = \mu_0$) and differentiable on nonzero scale resolutions ($\varepsilon \neq 0$). In these conditions the movement laws given by Equation (15), regardless of any coordinate transformations, can be integrated with the scale laws, regardless of any scale resolution transformations.

In such context, a further in-depth study of the systems dynamics needs to be done for the case in which the scale resolution is null, thus for the case in which Equation (15) defines a non-differential system. Let us consider that in Equation (15) the scale resolution in null, thus the restriction $\varepsilon \to 0$ is satisfied. Then Equation (15) takes the following form:

$$\frac{dX}{dT} = Y.$$

$$\frac{dY}{dT} = X - XZ. \tag{16}$$

$$\frac{dZ}{dT} = XY.$$

2.3. Motion Integration

The system in Equation (16) presents some interesting features, such as the phase space volume associated to the dynamics of the system being conserved, since it satisfies the relationship:

$$\frac{\partial}{\partial X}\left(\frac{\partial X}{\partial T}\right) + \frac{\partial}{\partial Y}\left(\frac{\partial Y}{\partial T}\right) + \frac{\partial}{\partial Z}\left(\frac{\partial Z}{\partial T}\right) = 0 \qquad (17)$$

This means that through the restriction $\varepsilon \to 0$, describing the transition from a differentiable Lorenz type system to a non-differentiable Lorenz type system, the system evolves from a dissipative one to a non-dissipative one. The non-differentiable Lorenz type system admits two integrals of the motions. The first is:

$$\frac{X^2}{2} - Z = k_1 = const. \qquad (18)$$

This relation is obtained by multiplying the first equation from (16) with Y and further substituting the term XY through the third equation of Equation (16) and finally integrating the result. The second movement's integral is:

$$\frac{1}{2}Y^2 - Z + \frac{1}{2}Z^2 = k_2 = const. \qquad (19)$$

Relation (19) is achieved by multiplying the second equation from Equation (16) with Y, substituting the terms XY and XYZ with the third equation from Equation (16), and finally integrating the result. Let us further note that the dynamics introduced through a non-differential Lorenz type system are described analytically through elliptic functions. In order to showcase this, let us consider the square of Equation (16) using Equations (18) and (19), written as:

$$\dot{X}^2 = Y^2 = 2k_2 + 2Z - Z^2 = 2k_2 + \left(X^2 - 2k_1\right) - \left(\frac{1}{2}X^2 - k_1\right)^2. \qquad (20)$$

It reveals the elliptic integral:

$$\int \frac{dX}{\sqrt{P(X)}} = \frac{1}{2}\int dT. \qquad (21)$$

with

$$P(X) = -X^4 + 4X^2(1 + k_1) + 4\left[2(k_2 - k_1) - k_1^2\right]. \qquad (22)$$

Admitting that the $P(X)$ polynom can be written as:

$$P(X) = \left(u_1^2 - X^2\right)\left(X^2 - u_2^2\right),$$

where u_1^2 and u_2^2 have the following expressions:

$$u_{1,2}^2 = 2\left[(1 + k_1) \pm (1 + 2k_2)^{1/2}\right]. \qquad (23)$$

The integral (21) becomes:

$$\int \frac{dX}{\left[\sqrt{\left(u_1^2 - X^2\right)\left(u_2^2 - X^2\right)}\right]} = \frac{i}{2}\int dT. \qquad (24)$$

Furthermore, by using the substitutions:

$$w = \frac{X}{u_2}, \quad s = \frac{u_2}{u_1} \qquad (25)$$

the integral (24) takes the Legendre form:

$$\frac{dw}{\sqrt{(1-w^2)(1-s^2w^2)}} = \frac{iu_1}{2}\int dT. \tag{26}$$

The solution to that integral is the Jacobi elliptic function sn,

$$w = sn\left[\frac{iu_1}{2}(\Delta T); s\right] \tag{27}$$

or in its original coordinates:

$$X = u_2 sn\left[\frac{iu_1}{2}(\Delta T); s\right]. \tag{28}$$

From here, through Equation (18) we will get the expression for Z:

$$Z = \frac{u_2^2}{2}sn^2\left[\frac{iu_1}{2}(\Delta T); s\right] - k_1. \tag{29}$$

while by using Equation (19) we can obtain the expression for Y:

$$Y_{1,2} = \pm\left\{2(k_2 - k_1) + k_1^2 + u_2^2(1 + k_1)sn^2\left[\frac{iu_1}{2}(\Delta T); s\right] - \frac{u_2^4}{4}sn^4\left[\frac{iu_1}{4}(\Delta T); s\right]\right\}^{1/2}. \tag{30}$$

Now, the expressions for X, Y and Z can be simplified even more if we take into account the relationship:

$$sn\left[\frac{iu_1}{2}(T - T_0); s\right] = i\frac{sn\left[\frac{iu_1}{2}(\Delta T); s'\right]}{cn\left[\frac{iu_1}{2}(\Delta T); s'\right]}. \tag{31}$$

cn is a Jacobi elliptical function, s is the modulus of the elliptical functions cn and sn and s' is the complementary modulus $s'^2 = 1 - s^2 a$ and $\Delta T = T - T_0$.

In the following we will attempt to implement the solution gained by Lorenz system to a well-known system covering the ejection and dynamics of particles.

3. Plasma Modelling

The fractal representation of phenomena like laser produced plasma has a simplified elegance, as it confines all the complex intricate behaviors in a handful of parameters. This is a great advantage when attempting to simulate a wide range of behaviors for multiple external conditions, spatial and temporal coordinates, etc. However, the interpretation of the obtained results requires a direct correspondence between the compact fractal parameters and real, measurable plasma parameters. Given the description of our three normalized functions (X, Y and Z) they define complex functions that will be used further to describe formation and evolution of the ejected particles at different scale resolutions. Therefore, in our paradigm: X will define the particle distributions, Y will define the charged particle current as a function and Z will define the charge density fluctuations induced by some unbalances of the transient electrical field during expansion, and can be associated with the ratio between the kinetic and thermal energy of the ejected particles. Let us note that we will operate with the previous functions and parameters as normalized quantities. Since the Laser Produced Plasmas (LPP) dynamics are described through continuous and non-differentiable curves (fractal curves with different degrees of fractality), we had to operate in this situation with what we call the singularity spectrum. If in the system the dynamics are characterized by only one fractal dimension then we are dealing with mono fractal dynamics. If for a system (which is the case for laser produced plasmas) we deal with simultaneous dynamics with various fractal dimensions, then the role of the singularity spectra is not only to showcase the variation domain of these dimensions but also the characteristic classes associated by defining the strange attractors.

In Figure 1a,b we have represented the particle velocity distribution at various scale resolutions and at various moments in time. We can see from the two representations that the evolution of the plasma in time, implies the presence of families of particles defined by specific scale resolutions and velocity distributions. With an increase of the scale resolution we see that the fractal image of the laser produced plasma showcases the appearance of a one particle distribution centered on a relatively high velocity, at low resolutions scale and a consistent increase number of distributions centered around lower velocities. The presence of multiple structures in laser produced plasma has been previously attested and reported as a direct result of the multiple ejection mechanism and plasma interaction with the background gas. We note that the presence of multiple ejection mechanisms [17–20] (Coulomb explosion, explosive boiling, phase explosion, etc.) will lead to the presence, within the plasma volume, of particles or plasma structure defined by different fractalization and scale resolutions. The simulations also showcase a transition from quasi mono-energetic particles ejected through a single ablation mechanism to an amalgam of particles with different energetic distributions specific to each ejection mechanism involved.

In Figure 1b we represented the particle distribution with the scale resolution for various moments in time. As the plasma evolves we see the presence of multiple distribution sites centered across low (s < 0.3) and high (0.4 < s < 1) resolution scales. This suggests that the first ejected particles ($\Delta T = 40$) are defined only by one scale resolution, characteristic of the Coulomb explosion mechanism. For longer period of time, when the thermal mechanism is dominant, we see multiple distributions centered on higher values of the scale resolutions. Therefore, our simulations capture the ejection of a series of particles with different kinetic energies and fractalizations. As such, we can now define a scale resolution for each of the ejection mechanisms (~ 0.2 for Coulomb Explosion, ~0.6 for Explosive Boiling and ~1 for the removal of complex structures) ensuring a wide covering of our theoretical model, transcending the Coulomb temporal scale, up to the explosive boiling ejection scenario and the ejection of clusters and more complex structures.

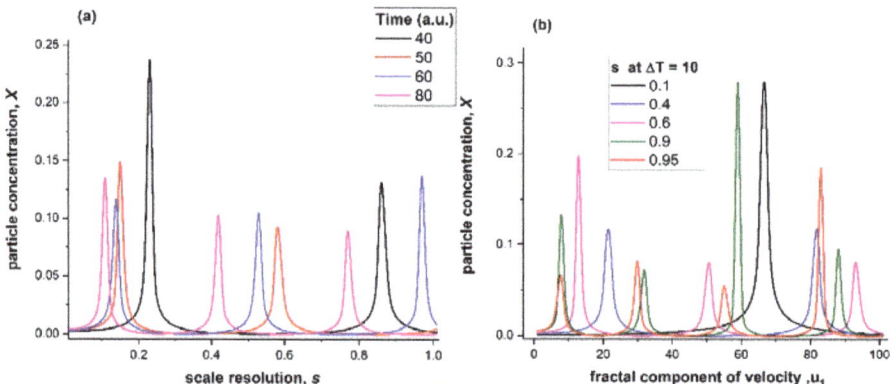

Figure 1. (a) Particle velocity distribution for various scale resolutions, (b) particle distribution with the scale resolution distribution at various moments in time.

In Figure 2a we have represented the charge density fluctuations induced by some unbalances with respect to the plasma volume (ξ). We can see that for short moments of time we observed a separation of the charges. The obtained distribution in the fractal space resembles the double layer distribution in the case of classical plasma physics [18]. As the time evolves, the double-layer-like distribution moves towards higher distances, and for longer moments of time a secondary one appears. This behavior is in line with the transient double layer scenario published by Bulgakova [20]. The charge separation that occurs during the Coulomb explosion will act as a driving force behind the expansion of the particles. The double layer formed during the particle expansion will lead to the acceleration of ions and the deceleration of the electron. This exchange will induce some current oscillations [21] with the frequency depending of the nature of the material, and thus on its fractality. The formation of the second plasma structure, expanding with a lower velocity and being described as having a different temperature [22], will also lead to the formation of a double layer between the fast structure and slow structure [20]. This phenomenon was showcased by us in Figure 2a, where we can see the appearance of a second signature of the double layer in the charged particle distribution at a later evolution time. The intensity of the function represented in Figure 2a increases with the increase of time and we can see that at longer expansion time, where the formation of a second plasma structure becomes visible, another more intense double layer forms. This result was expected, as the multiple structures formed within the plasma volume suffered a spatial and temporal expansion. In order to maintain the composed global shape and to compensate the loses, particle density, with an increase in the separation between the two structures, of the electrical field defining the double layers increase their amplitudes. In Figure 2b, we have represented the particle current evolution in time for various scale resolutions. We noticed that the particles present an oscillatory dynamic, as was previously reported by [23] and experimentally proven in [24] and [25]. The frequency changes with the scale resolution, thus we deduce that each component of the plasma might present a different oscillatory behavior dictated by the characteristics of the double layer formed in the area separating the structures.

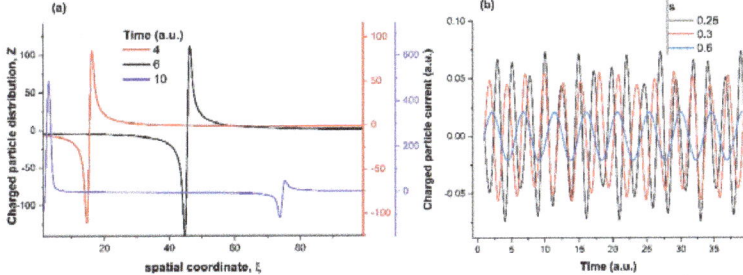

Figure 2. (a) Charged particles' distribution at various moments in time, (b) charge particle current induced by fluctuation current at the resolution scale.

4. Experimental Confirmation

In order to verify if our theoretical assumptions can have experimental correspondents we investigated by means of ICCD fast camera imaging and space-and-time-resolved optical emission spectroscopy of a plasma generated by an ns laser beam on a Chalcopyrite mineral target. The experiments were performed in fixed external conditions (laser fluence 5 J/cm^2 and background pressure of 10^{-2} Torr). The choice of the relatively simple mineral comes from its composition, having elements with different physical properties (S, Cu and Fe) which will allow a better showcase of phenomena like: particle separation, ionic oscillations and plume splitting. Further details on the experimental set-up can be found in [6,12].

In Figure 3a we have represented the spatial distribution of spectral region 420–430 nm after a time delay of 650 ns. This region is significative for our laser produced plasmas as it contains all the elements composing the target (Cu I-II, Fe I-II and S II). We noticed that the atomic emission lines can be found only for short distance while the ionic lines start and end emission at considerably longer distances. This result confirms other findings from literature and underlines the separation of the plasma components into fast and slow [26], respecting the ejection mechanism behind each species. In Figure 3b we have plotted the spatial distribution of representative emission lines for all the atomic and ionic species. Due to the difference in expansion velocities between ionic (S II-15.4 km/s,

Fe II-14.3 km/s and Cu II-11.4 km/s) and atomic species (Fe I-6.2 km/s and Cu I-4.4 km/s), the ions can be seen expanding at longer distances with respect to the target surface. The slightly increased velocity of S ions compared with Fe or Cu ones is a direct consequence of the acceleration in the double layer generated through Coulomb expansion [27]. We also notice periodic fluctuations (oscillations) on the spatial distribution of all the ionic species. The period of these oscillation are of approximately 900 kHz in good agreement with other reports of ionic oscillations in laser produced plasmas [21,24,28]. The first attempts for the comprehension of this "peculiar" behavior was based on the formation of single or multiple double-layers in the very vicinity of the target. This picture was the main focus to a long series of papers reporting on charge separation in laser-produced plasma, mainly from the 1980s [29,30]. Eliezer and Hora [23] gathered, in a very comprehensive manner, the state of the art regarding the double and multiple layers in laser-produced plasmas. One of the remarkable results reported are experimental proofs with double-layer electric fields of 105–106 V/cm and widths of 10–100 Debye lengths. In the past few years, three other theoretical approaches where proposed. One based on the fractal model developed as the interaction between two fractal structures [5,31], and their corresponding interface (generally, this interface delineates the double layer), with the second ones being on differential physics [28]: a collisional model based on the plasma ion frequency and electron-ion collision rate in the context of the Lieberman's model for plasma immersion ion implantation, and finally, one based on the AC Josephson effect. So, at this point there is no real consensus for the real mechanism behind the oscillatory behavior but intense theoretical and experimental work is undergoing to shed some light on it.

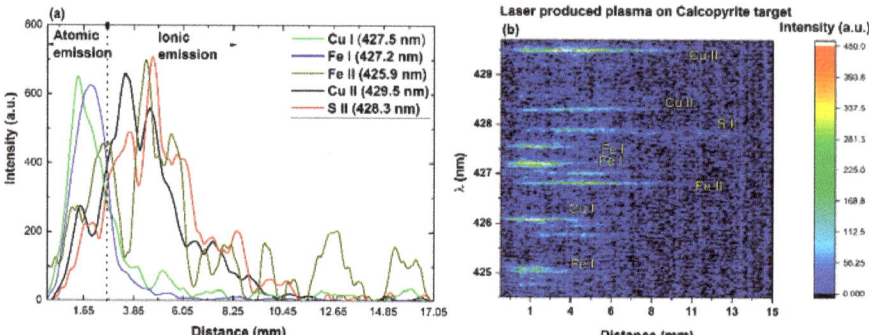

Figure 3. (a) Spatial mapping of Cu, S and Fe emission after 650 ns, (b) bi-dimensional representation of atomic and ionic emission.

To verify how the distribution of the compositional species within the plasma volume would affect the overall behavior of the laser produced plasma, we recorded the overall emission of the plasma at various moments in time. In Figure 4a we show a representative image of the plasma at a time-delay of 650 ns. If we perform cross section across the main expansion axis (centered around the orthogonal direction on the target in the impact point), three areas are more visible as the time delay increases. The first area seen at a larger distance corresponds to an ion rich area (confirmed by the data presented in Figure 3b) expanding with 16 km/s; the second one corresponds to an atomic rich area (~6 km/s); and the emission maxima corresponds to the maxima of the atomic emission. The last area has a small intensity and it is in the proximity of the target. It corresponds to the emission from atomic species colliding with clusters and microdroplets ejected from the target. This emission was not seen in the spectra's resolved measurements but remains a trademark of a strong thermal mechanism and the ejection of clusters and nanoparticles [32]. The difference in expansion velocity is a signature of the fundamental ejection mechanism behind each structure. The first structure contains mainly ions and it is induced during Coulomb explosion. The second structure contains mainly atomic species ejected by the thermal mechanism which require a longer incubation time (up to a few ns), thus displaying a lower velocity. The third structure mainly contains nanostructures, or clusters ejected directly from the target and expanding with a significantly lower velocity (generally found in the order of hundreds of m/s).

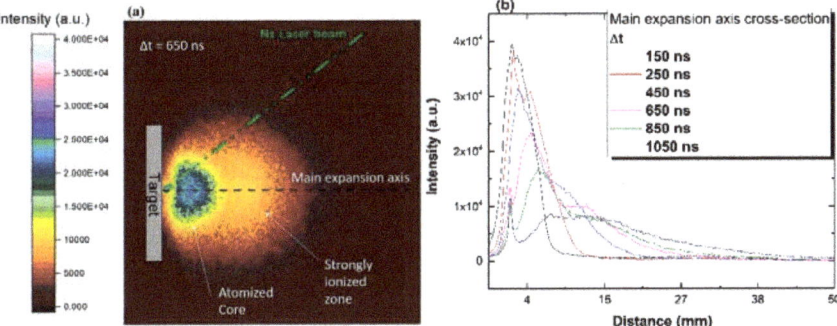

Figure 4. (a) ICCD fast camera image of a LPP Chalcopyrite collected after 650 ns, (b) cross section for a series of images extracted at various time delays.

The experimental data extracted from a plasma generated on chalcopyrite mineral by irradiation with an ns laser beam confirms the theoretical projections presented in Section 3. The temporal separation the plasma components based on their inner properties accurately reflects the separation based on the resolution scale of each type of ablation mechanism and the presence of multiple distribution on the scale resolution representation. One of the most important results is the prediction and confirmation of ionic oscillations by using invasive techniques. The oscillatory behavior coupled with heterogenic dynamics of different ions is well in line with the image depicted by other groups [20,33] and by our group in past papers [5,13,31]. In this paper the oscillatory behavior and the double layer-like distribution appear as natural solutions to the initial paradigm which translated the initial complex Lorenz system from laser to target to plasma. As such, we were able to build a robust theoretical model that can contains the set of parameters for the laser, target and the projection of the particle dynamic after ejection as solutions to the initial system.

5. Conclusions

A non-differential Lorenz system was built by projecting a differential Lorenz system on a fractal space. Simulations are performed for a wide range of scale resolutions showcasing the appearance of multiple distribution centered on different velocities attributed to the various plasma formed through different removal mechanisms. Current oscillations were also predicted as a result of the appearance of multiple double layers during expansion.

The theoretical simulations were confronted with experimental data extracted by means of ICCD fast camera imaging and space and time resolved optical emission spectroscopy of a complex plasma generated by ns-laser ablation on a chalcopyrite sample. Space and time resolved measurement revealed an oscillating behavior seen in the emission of the Cu, Fe and S ions. The ions were found to expand with various velocities specific to each species present in the plasma. The ICCD imaging revealed the split into two structures (fast and slow) expanding with different velocities. The values were found consistent with the ones of the individual species seen through spectrally resolved measurements. The experimental data is in good agreement with the major predictions made by the theoretical model based on the Lorenz system.

Author Contributions: Conceptualization, M.A. and S.A.I.; methodology, S.A.I., A.A. and M.A.; investigation, S.A.I., A.A and F.E; writing—original draft preparation, S.A.I. and M.A.; writing—review and editing, F.E. and A.A.; visualization, S.A.I.; supervision, M.A.

Funding: This work has been funded by the National Authority for Scientific Research and Innovation in the framework of the Nucleus Program—16N/2019.

Conflicts of Interest: The authors declare no conflict of interest. The funders had no role in the design of the study; in the collection, analyses, or interpretation of data; in the writing of the manuscript, or in the decision to publish the results.

References

1. Phipps, C.R. *Laser Ablation and Its Applications*; Springer Series in Optical Sciences; Springer: Boston, MA, USA, 2007; Volume 129, ISBN 978-0-387-30452-6.
2. Autrique, D.; Clair, G.; L'Hermite, D.; Alexiades, V.; Bogaerts, A.; Rethfeld, B. The role of mass removal mechanisms in the onset of ns-laser induced plasma formation. *J. Appl. Phys.* **2013**, *114*, 023301. [CrossRef]
3. Toftmann, B.; Schou, J. Time-resolved and integrated angular distributions of plume ions from silver at low and medium laser fluence. *Appl. Phys. A Mater. Sci. Process.* **2013**, *112*, 197–202. [CrossRef]
4. Geohegan, D.B.; Puretzeky, A.A.; Duscher, G.; Pennycook, S.J. Time-resolved imaging of gas phase nanoparticle synthesis by laser ablation. *Appl. Phys. Lett.* **1998**, *72*, 2987. [CrossRef]
5. Irimiciuc, S.A.; Mihaila, I.; Agop, M. Experimental and theoretical aspects of a laser produced plasma. *Phys. Plasmas* **2014**, *21*, 093509. [CrossRef]
6. Irimiciuc, S.A.; Bulai, G.; Gurlui, S.; Agop, M. On the separation of particle flow during pulse laser deposition of heterogeneous materials—A multi-fractal approach. *Powder Technol.* **2018**, *339*, 273–280. [CrossRef]
7. Anisimov, S.I.; Luk'yanchuk, B.S. Selected problems of laser ablation theory. *Phys.-Uspekhi* **2002**, *45*, 293–324. [CrossRef]
8. Bulgakova, N.M.; Stoian, R.; Rosenfeld, A.; Hertel, I.V.; Marine, W.; Campbell, E.E.B. A general continuum approach to describe fast electronic transport in pulsed laser irradiated materials: The problem of Coulomb explosion. *Appl. Phys. A Mater. Sci. Process.* **2005**, *81*, 345–356. [CrossRef]
9. Kelly, R.; Miotello, A. On the role of thermal processes in sputtering and composition changes due to ions or laser pulses. *Nucl. Instrum. Methods Phys. B* **1998**, *141*, 49–60. [CrossRef]
10. Bulgakova, N.M.; Stoian, R.; Rosenfeld, A.; Hertel, I.V. Continuum Models of Ultrashort Pulsed Laser Ablation. In *Laser-Surface Interactions for New Materials Production*; Miotello, A., Ossi, P., Eds.; Springer: Berlin/Heidelberg, Germany, 2010; Volume 130, pp. 81–97.
11. Autrique, D.; Chen, Z.; Alexiades, V.; Bogaerts, A.; Rethfeld, B. A multiphase model for pulsed ns-laser ablation of copper in an ambient gas. *AIP Conf. Proc.* **2012**, *1464*, 648–659.
12. Irimiciuc, S.; Bulai, G.; Agop, M.; Gurlui, S. Influence of laser-produced plasma parameters on the deposition process: In situ space- and time-resolved optical emission spectroscopy and fractal modeling approach. *Appl. Phys. A Mater. Sci. Process.* **2018**, *615*, 1–14. [CrossRef]
13. Irimiciuc, S.A.; Gurlui, S.; Nica, P.; Focsa, C.; Agop, M. A compact non-differential approach for modeling laser ablation plasma dynamics. *J. Appl. Phys.* **2017**, *121*, 083301. [CrossRef]
14. Merches, I.; Agop, M. *Differentiability and Fractality in Dynamics of Physical Systems*; World Scientific: Singapore, 2015; ISBN 978-981-4678-38-4.
15. Haken, H. *Synergetics*; Springer: Berlin, Germany, 1983; ISBN 978-3-642-88338-5.
16. Arfken, G.; Weber, H.; Harris, F.E. *Mathematical Methods for Physicsists*, 7th ed.; Academic Press: Cambridge, MA, USA, 2012; ISBN 9780123846556.
17. Dachraoui, H.; Husinsky, W.; Betz, G. Ultra-short laser ablation of metals and semiconductors: Evidence of ultra-fast Coulomb explosion. *Appl. Phys. A Mater. Sci. Process.* **2006**, *83*, 333–336. [CrossRef]
18. Kelly, R.; Miotello, A. Comments on explosive mechanisms of laser sputtering. *Appl. Surf. Sci.* **1996**, *96–98*, 205–215. [CrossRef]
19. Merino, M.; Ahedo, E. Two-dimensional quasi-double-layers in two-electron-temperature, current-free plasmas. *Phys. Plasmas* **2013**, *20*, 023502. [CrossRef]
20. Bulgakov, V.; Bulgakova, N.M. Dynamics of laser-induced plume expansion into an ambient gas during film deposition. *J. Phys. D Appl. Phys.* **1999**, *28*, 1710–1718. [CrossRef]

21. Focsa, C.; Gurlui, S.; Nica, P.; Agop, M.; Ziskind, M. Plume splitting and oscillatory behavior in transient plasmas generated by high-fluence laser ablation in vacuum. *Appl. Surf. Sci.* **2017**, *424*, 299–309. [CrossRef]
22. Jiang, L.; Tsai, H.-L. A plasma model combined with an improved two-temperature equation for ultrafast laser ablation of dielectrics. *J. Appl. Phys.* **2008**, *104*, 093101. [CrossRef]
23. Eliezer, S. Double layers in laser-produced plasmas. *Phys. Rep.* **1989**, *172*, 339–407. [CrossRef]
24. Singh, S.C.; Fallon, C.; Hayden, P.; Mujawar, M.; Yeates, P.; Costello, J.T. Ion flux enhancements and oscillations in spatially confined laser produced aluminum plasmas. *Phys. Plasmas* **2014**, *21*, 093113. [CrossRef]
25. Borowitz, J.L.; Eliezer, S.; Gazit, Y.; Givon, M.; Jackel, S.; Ludmirsky, A.; Salzmann, D.; Yarkoni, E.; Zigler, A.; Arad, B. Temporally resolved target potential measurements in laser-target interactions. *J. Phys. D Appl. Phys.* **1987**, *20*, 210–214. [CrossRef]
26. Irimiciuc, S.; Boidin, R.; Bulai, G.; Gurlui, S.; Nemec, P.; Nazabal, V.; Focsa, C. Laser ablation of (GeSe2)100−x(Sb2Se3)$_x$ chalcogenide glasses: Influence of the target composition on the plasma plume dynamics. *Appl. Surf. Sci.* **2017**, *418*, 594–600. [CrossRef]
27. Stoian, R.; Ashkenasi, D.; Rosenfeld, A.; Campbell, E.E.B. Coulomb explosion in ultrashort pulsed laser ablation of Al_2O_3. *Phys. Rev. B* **2000**, *62*, 13167–13173. [CrossRef]
28. Nica, P.; Agop, M.; Gurlui, S.; Focsa, C. Oscillatory Langmuir probe ion current in laser-produced plasma expansion. *EPL* **2010**, *89*, 65001. [CrossRef]
29. Ludmirsky, A.; Givon, M.; Eliezer, S.; Gazit, Y.; Jackel, S.; Krumbein, A.; Szichman, H. Electro-optical measurements of high potentials in laser produced plasmas with fast time resolution. *Laser Part. Beams* **1984**, *2*, 245–250. [CrossRef]
30. Ludmirsky, A.; Eliezer, S.; Arad, B.; Borowitz, A.; Gazit, Y.; Jackel, S.; Krumbein, A.D.; Salzmann, D.; Szichman, H. Experimental Evidence of Charge Separation (Double Layer) in Laser-Produced Plasmas. *IEEE Trans. Plasma Sci.* **1985**, *13*, 132–134. [CrossRef]
31. Irimiciuc, S.A.; Agop, M.; Nica, P.; Gurlui, S.; Mihaileanu, D.; Toma, S.; Focsa, C. Dispersive effects in laser ablation plasmas. *Jpn. J. Appl. Phys.* **2014**, *53*, 116202. [CrossRef]
32. Boulmer-Leborgne, C.; Benzerga, R.; Perrière, J. Nanoparticle Formation by Femtosecond Laser Ablation. *J. Appl. Phys D* **2007**, *40*, 125–140.
33. Babushok, V.I.; DeLucia, F.C.; Gottfried, J.L.; Munson, C.A.; Miziolek, A.W. Double pulse laser ablation and plasma: Laser induced breakdown spectroscopy signal enhancement. *Spectrochim. Acta Part B At. Spectrosc.* **2006**, *61*, 999–1014. [CrossRef]

© 2019 by the authors. Licensee MDPI, Basel, Switzerland. This article is an open access article distributed under the terms and conditions of the Creative Commons Attribution (CC BY) license (http://creativecommons.org/licenses/by/4.0/).

Article

Charged Particle Oscillations in Transient Plasmas Generated by Nanosecond Laser Ablation on Mg Target

Maricel Agop [1,2], Ilarion Mihaila [3], Florin Nedeff [4] and Stefan Andrei Irimiciuc [5,*]

1. Department of Physics, "Gh. Asachi" Technical University of Iasi, 700050 Iasi, Romania; m.agop@tuiasi.ro
2. Romanian Scientists Academy, 54 Splaiul Independentei, 050094 Bucharest, Romania
3. Integrated Center for Studies in Environmental Science for North-East Region (CERNESIM), Alexandru Ioan Cuza University of Iasi, 700506 Iasi, Romania; ilarion.mihaila@uaic.ro
4. Mechanical Engineering, "Vasile Alecsandri" University of Bacau, Calea Mărășești, 600115 Bacău, Romania; florin_nedeff@ub.ro
5. National Institute for Laser, Plasma and Radiation Physics, 409 Atomistilor Street, 077125 Bucharest, Romania
* Correspondence: stefan.irimiciuc@inflpr.ro

Received: 12 January 2020; Accepted: 7 February 2020; Published: 17 February 2020

Abstract: The dynamics of a transient plasma generated by laser ablation on a Mg target was investigated by means of the Langmuir probe method and fractal analysis. The empirical data showcased the presence of an oscillatory behavior at short expansion times (<1 µs) characterized by two oscillation frequencies and a classical behavior for longer evolution times. Space- and time-resolved analysis was implemented in order to determine main plasma parameters like the electron temperature, plasma potential, or charged particle density. In the motion fractal paradigm, a theoretical model was built for the description of laser-produced plasma dynamics expressed through fractal-type equations. The calibration of such dynamics was performed through a fractal-type tunneling effect for physical systems with spontaneous symmetry breaking. This allows both the self-structuring of laser-produced plasma in two structures based on its separation on different oscillation modes and the determination of some characteristics involved in the self-structuring process. The mutual conditionings between the two structures are given as joint invariant functions on the action of two isomorph groups of SL(2R) type through the Stoler-type transformation, explicitly given through amplitude self-modulation.

Keywords: laser ablation; charged particle oscillations; Langmuir probe; fractal analysis; Lie groups; joint invariant functions

1. Introduction

The fundamentals of laser–matter interactions have been at the core of developing a wide range of applications such as pulsed-laser deposition, material processing, or even medical applications. In the past few years, a great deal of interest has been given to fundamental aspects of high-power laser–matter interaction, especially in the context of the development of unique infrastructures that can generate novel applications. Pulsed-laser deposition has gained a lot of attention in the past 10 years as one of the best techniques to produce complex films with relatively complicated stoichiometry [1–3]. The technique has a proven flexibility in terms of the deposition geometry [4] and target or background gas nature [5,6]. Great advancements have been made toward understanding the fundamental aspects of laser ablation for a better control of the thin-film deposition technique.

In the past years, the focus has also been on high-power laser interaction with materials, especially in the context of the development of a new generation of high-power lasers and their potential applicability in technological development and nuclear physics. Regardless of the application aims or the fluence regime used, there are several diagnostic techniques that can showcase the fundamental dynamics and behaviors of laser-produced plasmas. Over time, a wide range of investigation techniques have been developed like optical emission spectroscopy [7–9], mass spectrometry [10], or Langmuir probes [11–14] (LP), and these have been implemented to highlight phenomena like plasma structuring [15,16], molecule formation [17,18], elemental distribution in complex plasmas, and ejected particle behavior in various conditions. Out of all the techniques presented in the literature, LPs have shown great versatility being implemented for a wide range of materials [11,19] and various irradiation conditions [10,20]. The LP method presents itself as a relatively simple diagnostic tool for plasma investigations, consisting of submerging a metallic electrode (of cylindrical, plane, or spherical geometry) in the plasma in order to record the ionic and electronic current or a mixture of the two currents, as selected by the applied voltage. The technique was first developed for steady-state discharge plasmas presenting local or global thermodynamic equilibrium. Laser-produced plasma (LPP) has a transient nature, with all its parameters presenting a complex spatial distribution and temporal dependence and being largely highly directional. The LP theory has been adapted for LPP by sampling the charged particle temporal traces and considering, besides the thermal movement, the drift movement of the ejected particle [10,11]. LP techniques offer insight into the ejected particle dynamics from a small plasma volume and, thus, can be implemented for axial and angular measurements, and it can easily be adapted for a wide range of diagnostics geometries. In the time-resolved approach to LP diagnostics, it can offer information from a wide set of plasma properties (electron temperature, plasma potential, charged particle density, collisions frequency, etc.) during the plasma expansion.

The flexibility of the technique allowed for some exciting findings. Charged particle oscillations were evidenced with the use of a single cylindrical probe [21], heated probe [12], and multiple probe [22,23] configurations. Their nature is often debated with some reports presenting their roots in the dynamics of a plasma structure generated by electrostatic mechanisms [24], or induced by the transient double layers generated through plasma structuring [25]. Other novel theoretical approaches have been presented either in the framework of a fractal theoretical model [12,26] or built around Lorenz-type systems [27]. The complete behavior of plasma charged particles can be accounted for in the framework of the aforementioned fractal paradigm with the oscillatory behavior being explained through the presence of dissipative [12] or dispersive [26] effects, and the classical behavior is better showcased in the compact fractal hydrodynamic model [28]. The fractal paradigm proposes that every dynamic variable describing laser ablation plasma systems acts as the limit of families of functions. These functions can be differentiable for a non-zero scale resolution and non-differentiable for a null-scale resolution. The method is well-adapted for LPP dynamics, where the analyses conducted at a finite resolution scale imply the development of a new theoretical paradigm and new geometries. The motion curves can be seen as geodesics in the fractal space and forming a fractal fluid, where the ejected particles (the entities of the fluid) are replaced by their geodesics.

In this paper, we report on some novel results obtained by implementing the electrical diagnosis of laser ablation plasma produced on a Mg target. The Langmuir probe method was used to describe the behavior of laser-produced plasmas generated at different laser fluences and different measurement distances. A non-differentiable theoretical model based on a fractal interpretation of laser-produced plasma dynamics was used. The model is built around the interactions between the multiple plasma structures generated by laser ablation. In such contexts, the joint invariant functions on the action of two groups of SL(2R) type are given as mutual conditionings between the plasma substructures.

2. Experimental Setup

A schematic view of the experimental setup is presented in Figure 1. The experiments have been performed in a stainless-steel vacuum chamber pumped down to 2×10^{-5} Torr residual pressure. The radiation from a Quantel Brilliant, Nd-YAG nanosecond laser (355 nm-3rd harmonic, pulse width = 5 ns, 10 Hz repetition frequency, variable fluences) was focused by a f = 30 cm lens onto a magnesium target (the spot diameter at the impact point was approximately 0.3 mm) placed in the vacuum chamber. The magnesium target rotated during the experiments and was electrically grounded from the vacuum chamber. Before each measurement, a surface-cleaning procedure was implemented in order to remove the oxide layer present at the surface of the target.

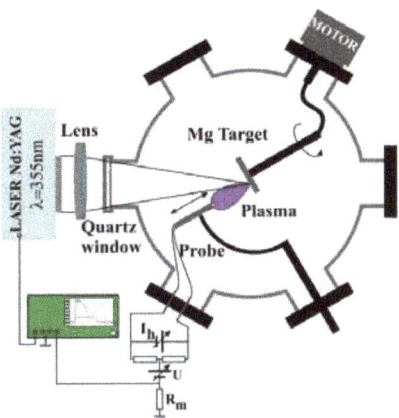

Figure 1. Experimental setup.

The ionic and electronic currents were collected from the ablation plasma plume using a tungsten-heated cylindrical Langmuir probe with a 0.25 mm diameter and 3 mm length. The probe was heated in order to avoid deposition of the target material on the probe, thus keeping constant the collecting area of the probe during measurements. In order to explore the dynamics of the expanding plasma, the probe was placed at different distances (1.5, 2, 2.5, and 3 cm) with respect to the target, in the maximum expanding direction of the plume. The Langmuir probe was also biased by voltages on a range of values between −5 and +10 V with a stabilized dc power source. Transitory signals were recorded by a digital oscilloscope and transferred to a PC for further analysis.

3. Langmuir Probe Measurements

For a bias of ±5 V applied on the Langmuir Probe, we are able to collect the saturation ionic and electronic currents, displayed in Figure 2a,b, for two different fluences characterizing 2.45 mm^3 plasma volume at 2 cm with respect to the target surface. By recording the saturation charge currents, we are assuring the collection of global ionic and electronic charge ejected as a result of the ablation process coupled with the ionization and neutralization process occurring during expansion of the laser-produced plasmas. We can distinguish different features for both saturation currents. The electronic current has a longer lifetime induced by the increased collision rate and the higher diffusion rate of the electrons as opposed to the ions. For the same applied voltage, the ionic Mg species have a lifetime of 3 μs while the electrons present almost twice as much, approximately 6 μs. There is, however, a recurring oscillating feature that can be found on all collected signals regardless of the applied voltage. The choice to represent here only the temporal trace for the saturation currents is supported by their higher amplitude and a better highlight of the smaller features, like the oscillatory part of the signal, noticeable below 1 μs. In the inset of Figure 2a,b we can see a zoomed in view of the oscillatory regime. We observe that for higher fluences, the first oscillatory maximum is reached at a shorter expansion time, with no significant increase in the amplitude of the current. The results indicate an increase in the kinetic energy of the ejected particle as induced by the increase in laser fluence, while the overall ejected charge remains quasi-constant. The oscillatory behavior has been previously reported for a wide range of materials, in classical or more complex ablation geometries [12,21–23].

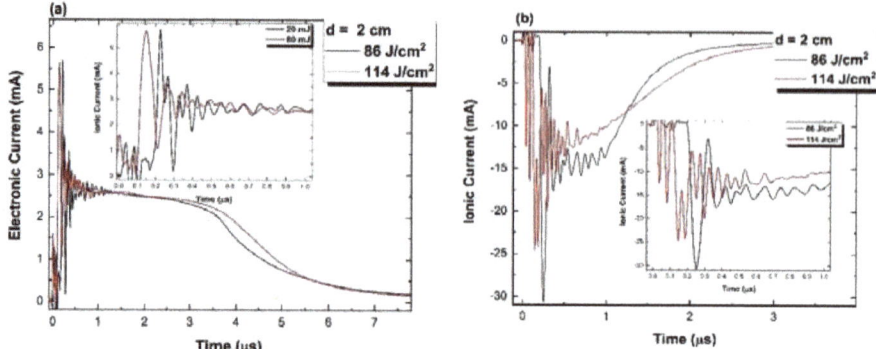

Figure 2. Temporal traces of the electronic (**a**) and ionic (**b**) current collected 2 cm from the target.

The nature of these oscillations is accepted, in some differentiable theoretical models, to be induced by the transient double layer forming as a result of the electrostatic ejection mechanism. This assertion would mean that the charged particles will oscillate with a frequency induced by the experimental conditions, as previously reported by our group in [11,12,28]. By performing fast Fourier transform (FFT) analysis on the collected signals, we observed two oscillation frequencies. These values showcase the strong dependence on the laser fluence (Figure 3a,b) and measurement distance (Figure 3c). The first oscillation frequency is of the order of tens of MHz (ranging from 15 up to 20 MHz), while the second one is of the order of a few MHz (ranging from 1 to 10 MHz). The results are in good agreement with our previous reports on Ni LPP, where we found values ranging from 5 up to 20 MHz or with those reported on a wider range of metals reported in [25]. The second oscillation frequency is induced by the splitting of the laser-produced plasma during expansion into two plasma structures expanding with different velocities and the appearance of a secondary transient double layer, which will accelerate the slower plasma structure [29,30]. Each plasma structure is characterized by a unique oscillation frequency. With the increase in the laser fluence, we notice an increase in the oscillation frequency (5%–50% depending on the measurement distance), which indicates the generation of a stronger electrical field generated through Coulomb explosion, followed by saturation for fluences higher than 80 J/cm^2. The observed oscillations are damped after approximately 1 µs. This can also be seen from the oscillation frequency evolution with the measurement distance where we notice a decrease of about 25% for both measured frequencies. Admittedly, the laser fluence values are much higher than those used in the pulsed-laser deposition or material processing applications, where we can find reports of fluences below 5 J/cm^2 [1–6], depending on the irradiation conditions and the nature of the thin film envisioned in each report. However, the results become relevant for the fundamental new generation of high-power laser–matter interactions.

The Langmuir probe theory is reportedly limited only for longer expansion times where the theoretical assumptions are met. For transient plasma and laser-produced plasma in general, the temporal evolution of some of the parameters (T_e, V_p, n_i) is determined by evaluating all the recorded I–V characteristics, in the hypothesis that in the moments of time selected for analysis, the plasma has properties imposed by LP theory. This hypothesis cannot always be verified, as the implementation of the technique induces some limitations. In the incipient part of the evolution (<1 µs) and in the proximity of the target (few mm), the probe-collecting surface is significantly larger than the measured plasma volume; therefore, LP theory is no longer valid, as the measurement electrode must not impact the plasma around it. In these special space–time coordinates, data reveals the presence of a complex oscillatory regime of the ionic and electronic temporal traces, which are also part of our reported results in this paper [11].

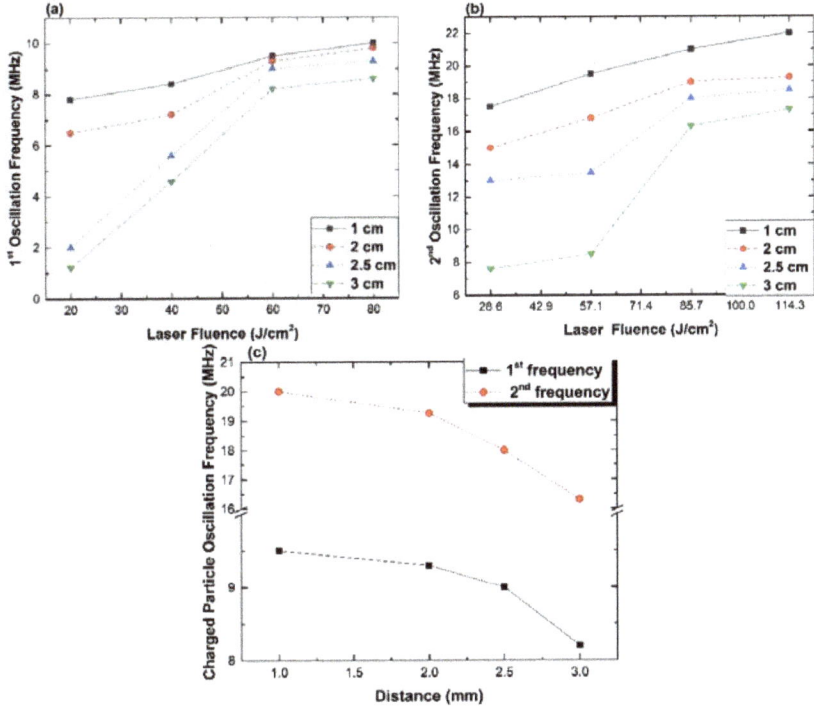

Figure 3. Oscillation frequency evolution with laser fluence (**a**,**b**) and measurement distance (**c**).

This allows us to characterize the temporal evolution of some plasma parameters like electron temperature, plasma potential, or ionic density. In order to achieve that, we implemented the methodology used in [11]. Briefly, consecutive bias potentials (in 50 points) ranging from −5 to +10 V were applied on the heated probe, and the plasma currents were collected from a well-defined plasma volume. The probe was moved at various distances with respect to the target's surface. An example is given in Figure 4. As anticipated in the previous paragraphs, all signals follow a similar pattern with an oscillatory part for short evolution times followed by a classical decreasing trend.

For a time-span of 6 μs using a step of 1 μs, we reconstruct the I–V characteristics for each moment in time. In Figure 5, we have presented representative characteristics describing the LPP at various distances after 1 μs (Figure 5a) and, for a fixed distance of 2.5 cm, the reconstructed characteristics at different moments in time (Figure 5b). We notice that for an instant temporal sequence, the axial dependence of the I–V characteristics is not significant; however, there is a substantial shift toward positive floating potential. This is due to the spatial distribution of the two sets of charges during expansion. The electrons generally have a more uniform distribution in the plasma; however, in light of the electrostatic ejection mechanism, the first electrons ejected will spatially occupy the position in front of the ejected ions. Although during expansion, there are other phenomena that need to be considered like ion neutralization, secondary ionizations, or molecule formation [18], this shift in the floating potential showcases the change in the ion-to-electron ratio within the plasma volume for a fixed moment in time (1 μs). The spatiotemporal evolution of all main plasma parameters considered here (electron temperature and plasma potential) follow a classical quasi-exponential decrease. To determine the aforementioned plasma parameters, we treated the I–V characteristics following the procedure from [10,11]. Briefly, by applying a logarithm procedure to the I–V characteristic curve, we will obtain a distribution that is defined by a linear increase, an inflection point, and a saturation region. The slope of that linear increase will define the electron temperature while the inflection point will be the plasma potential (Figure 5e). Details on this procedure can also be seen in [10–12].

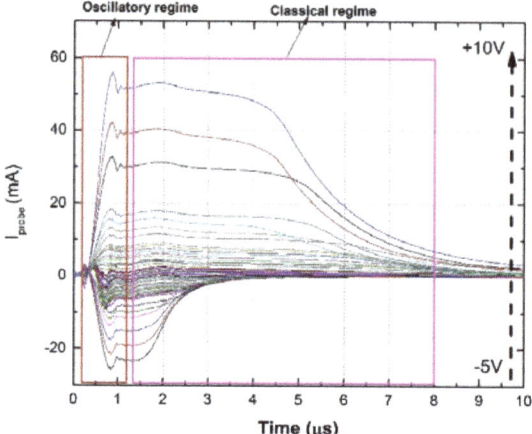

Figure 4. Temporal traces of the ionic and electronic currents recorded for a laser fluence of 56 J/cm^2 at 2.5 cm from the target.

The increase in the laser fluence leads to higher electron temperatures and plasma potentials (Figure 5c,d). The highest values found here are $T_e = 1$ eV and $V_p = 9.3$ V for the measurements performed at 1.5 cm and a laser fluence of approximately 114 J/cm^2, while the lowest are found for 28 J/cm^2 with the values decreasing with almost one order of magnitude: $T_e = 0.1$ eV and $V_p = 2$ V. For a fixed distance, the time-resolved analysis reveals important changes in both shapes of the I–V characteristics. These changes are induced by the decrease in all the plasma parameters as indicators of the laser-produced plasma expansion, particle density, and particle energy losses, showcased in the inset of Figure 5b where the temporal evolutions of T_e and V_p are presented.

The influence of the laser fluence over some plasma parameters is generally known. A higher laser fluence usually leads to the ejection of a higher density of particles with a higher kinetic and thermal energy. For our conditions, we synthesized the data in Figure 6a where we present the evolution of the ionic density with the laser fluence at various distances, and in Figure 6b, where the evolution of the ion drift velocity is presented. The results are in line with other reports where the same increase followed by a saturation regime can be seen. The ion drift velocities were determined by plotting the evolution of the ionic current maxima as a function of space and time. The slope of that representation defines the drift velocity.

The overall effect of the laser fluence depicted in Figures 3 and 6 is to enhance the oscillatory movement of the charged particles, increase the current densities, and, overall, increase the kinetic movement of the plasma. The kinetic enhancement of the LPP reaches a saturation regime, however, where neither the particle density nor expansion velocities increase. This could mean that the energy is lost on expelling large structures (clusters and nanoparticles). The result is in line with our previous results reported in [12] where the presence of a third plasma structure was seen in the particle velocity distribution at high laser fluence. The interaction between the two plasma structures and their unique signatures in the ionic and electronic saturation currents is still to be understood. The subtle difference in plasma plume dynamics at high fluence as opposed to the usual lower values used in applications such as PLD still needs to be investigated by other experimental techniques and cemented through comprehensive theoretical modeling.

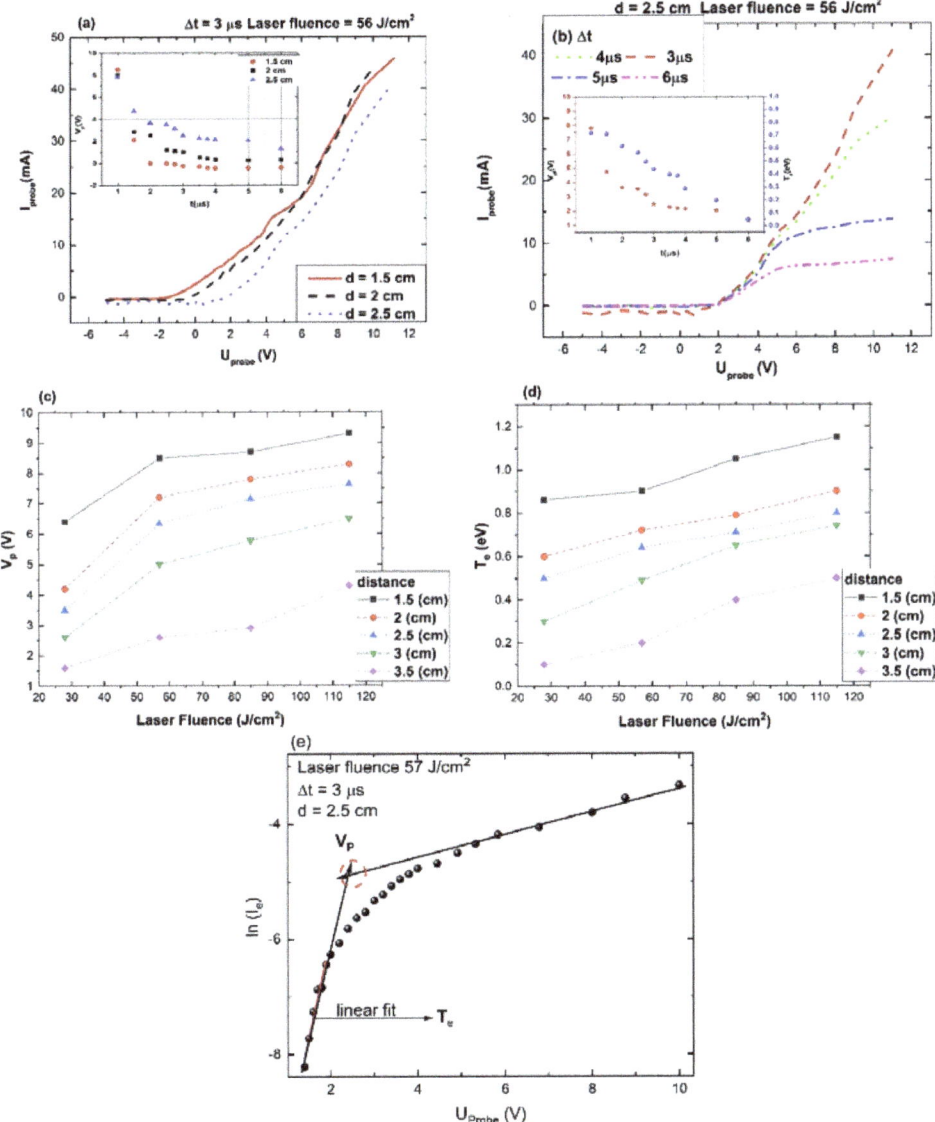

Figure 5. Evolution of the I–V characteristics in time (**a**) and space (**b**); plasma potential (**c**) and electron temperature (**d**) evolution with distance and laser fluence and an example of the logarithmic representation of the I–V characteristic (**e**).

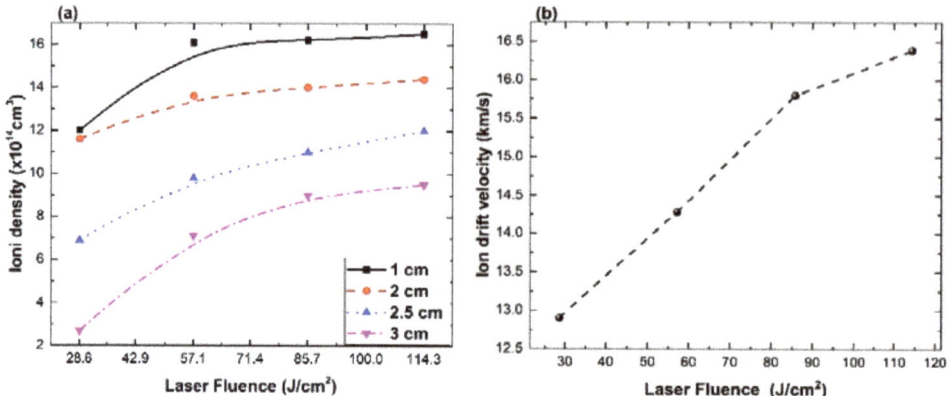

Figure 6. Ionic density (**a**) and velocity (**b**) evolution with the laser fluence.

4. Mathematical Model

4.1. Ablation Plasma as a Fractal Medium

Ablation plasma behaves like a fractal medium induced by the collisions process between its entities (electrons, ions, neutrals), with the results of the interactions taking the form of excitations, ionizations, or recombination [12,26,27]. Such an assumption can be theoretically sustained by this typical example: Between two successive collisions, the trajectory of the plasma particles is a straight line that becomes non-differentiable at the impact point. Considering now that all the collision impact points form an uncountable set of points, the trajectories of plasma particles become continuous but nondifferentiable curves, i.e., a fractal. Some follow-up clarifications from a mathematical and physical point of view need to be made: (i) If the laser-produced plasma expands at a lower background pressure (a lack of collisions between the ejected particles), it mimics a pre-fractal and approximates a fractal-sparse set [31]; (ii) if the ablation plasma expands at higher background pressures (an expansion rich in collisions), it mimics a fractal-dense set.

In such a context, fractal theories of motion [31–34] become functional for describing the dynamics of ablation plasma. Significant advancements for the implementation of a fractal paradigm to a classical physical system has been done in a series of papers [35–37], especially for the problem of the Schrodinger representation of fractal type [38–41]. The fundamental hypothesis of our model is that the dynamics of all ablation plasma entities can be characterized by fractal motion curves. These continuous but non-differentiable curves display the property of self-similarity at every point, which can be translated into a property of holography (every part reflects the whole). We can further use "holographic implementations" of the dynamics of any ablation plasma entity, for example, through special fractal-type fractal "regimes" (i.e., describing dynamics by using fractal-type equations at various scale resolutions).

The merit of developing theoretical models based on a fractal paradigm has been shown consistently by our group in the past 12 years. Our approach comes as an alternative to classical approaches in an attempt to offer a general model that can be adapted for various specific processes that need to be investigated. The model is complex and uses non-differentiable mathematics and is strongly linked to concepts from fractal physics. However, classical approaches to describe laser-produced plasma dynamics present some limitations. Theoretical models used to describe the laser ablation plasmas define either the first moments of LPP formation and one-dimensional expansion that occurs for small distances and short evolution times (dominated by laser-target interactions and fundamental particle removal mechanisms), or the later stages of expansion (a few μm after the laser pulse, starting from a few mm from the target), where the LPP has a three-dimensional expansion that is often described using classical hydrodynamics. Due to these major discrepancies in the space–time interaction and expansion scales, significant complications arise in developing a unique model that can cover all the complex phenomena involved in laser ablation. One of the domains of theoretical physics that has, at its core, the scale dependences of dynamic parameters is scale relativity theory (SRT) [31–34]. In SRT, the complete set of equations required to describe the dynamics of certain systems needs to contain the movement equations and those related to the non-differentiable processes. The non-differentiability indicates that for a plasma medium, the plasma entity evolution is defined by fractal curves (moving in a straight line between two collisions, and becoming nondifferentiable at the impact point) with the geodesics of fractal entities replacing the dynamics of plasma entities. In such a context, we can approximate the ablation plasma plume dynamics with that of a fractal fluid without interactions. The fractal hydrodynamic model [23] has been used in our group in recent years to describe the self-structuring of the ablation cloud during expansion and the evolution of the plasma parameters [27].

With respect to connections between the plasma parameters and those defined by the model, over time, we reported some clear relations. In [28], we give direct relations between the electron temperature and the fractal potential, ion density, normalized fractal entities density, electron thermal velocity, and non-differentiable velocity, and between the Debye length and the normalized specific length. In [11], we study the relationship between the fractalization degree given in the model and the heterogeneity of the ablated cloud in terms of particle sizes and properties.

4.2. Scale Covariant Derivative and Geodesics Equations

Let us now consider the scale covariance principle (the physical laws applied to the dynamics of the laser-produced plasmas are invariant with respect to scale resolution transformations [31–34]) and postulate that the transition from standard (differentiable) plasma physics to fractal (non-differentiable) plasma physics can be implemented by replacing the standard time derivative d/dt by the non-differentiable operator $\hat{d}/dt$ [35–37]:

$$\frac{\hat{d}}{dt} = \partial_t + \hat{V}^l \partial_l + \frac{1}{4}(dt)^{(\frac{2}{D_F})-1} D^{lp} \partial_l \partial_p, \qquad (1)$$

where

$$\begin{aligned}
\hat{V}^l &= V_D^l - V_F^l, \\
D^{lp} &= d^{lp} - i\overline{d}^{lp}, \\
d^{lp} &= \lambda_+^l \lambda_+^p - \lambda_-^l \lambda_-^p, \\
\overline{d}^{lp} &= \lambda_+^l \lambda_+^p + \lambda_-^l \lambda_-^p,
\end{aligned} \qquad (2)$$

$\partial_t = \frac{\partial}{\partial t}$, $\partial_l = \frac{\partial}{\partial X^l}$, $\partial_l \partial_p = \frac{\partial}{\partial X^l} \frac{\partial}{\partial X^p}$, $i = \sqrt{-1}$, $l, p = 1, 2, 3$.

In the above relations, $\hat{V}^l$ is the complex velocity, V_D^l is the differentiable velocity independent of the scale resolution dt, V_F^l is the nondifferentiable velocity dependent on the scale resolution, X^l is the fractal spatial coordinate, t is the non-fractal time with the role of an affine parameter of the motion curves, D^{lp} is the constant tensor associated with the differentiable-non-differentiable transition, λ_+^l is the constant vector associated with the forward differentiable-nondifferentiable physical processes, λ_-^l is the constant vector associated with the backward differentiable-nondifferentiable physical processes, and D_F the fractal dimension of the movement curve. For the fractal dimension, we can choose any definition, for example, the Kolmogorov-type fractal dimension or Hausdorff–Besikovici-type fractal dimension [42]. However, once chosen and becoming operational, it needs to be constant and arbitrary: $D_F < 2$ for the correlative physical processes; $D_F > 2$ for the non-correlative physical processes [31–34].

Now, the non-differentiable operator plays the role of the scale covariant derivative; namely, it is used to write the fundamental equations of ablation plasma dynamics in the same form as in the classic (differentiable) case. Under these conditions, accepting the functionality of the scale covariant principle, i.e., applying the scale covariant derivative, Equation (1), to the complex velocity field, Equation (2), in the absence of any external constraint, the geodesics equation of the ablation plasma takes the following form [32–34]:

$$\frac{d\hat{V}^i}{dt} = \partial_t \hat{V}^i + \hat{V}^l \partial_l \hat{V}^i + \frac{1}{4}(dt)^{(\frac{2}{D_F})-1} D^{lk} \partial_l \partial_k \hat{V}^i = 0. \tag{3}$$

This means that the fractal local acceleration $\partial_t \hat{V}^i$, the fractal convection $\hat{V}^l \partial_l \hat{V}^i$, and the fractal dissipation $D^{lk}\partial_l\partial_k \hat{V}^i$ of any ablation plasma entity balance themselves at any point of the motion fractal curve. Moreover, the presence of the complex coefficient of viscosity type $4^{-1}(dt)^{(\frac{2}{D_F})-1}D^{lk}$ in the laser-produced plasma (LPP) dynamics specifies that it is a rheological medium. Therefore, the plasma structures have memory, as a datum, by their own structure.

If the fractalization in the dynamics of LPP is achieved by Markov-type stochastic processes, which involve Lévy-type movements [31,42] of the plasma entities,

$$\lambda_+^i \lambda_+^l = \lambda_-^i \lambda_-^l = 2\lambda \delta^{il}, \tag{4}$$

where λ is a coefficient associated with the differentiable-nondifferentiable transition and δ^{il} is Kronecker's pseudo-tensor.

Under these conditions, the geodesics equation takes the following simple form:

$$\frac{d\hat{V}^i}{dt} = \partial_t \hat{V}^i + \hat{V}^l \partial_l \hat{V}^i - i\lambda(dt)^{(\frac{2}{D_F})-1} \partial^l \partial_l \hat{V}^i = 0. \tag{5}$$

For irrotational motions of the ablation plasma entities, the complex velocity field $\hat{V}^i$ takes the following form:

$$\hat{V}^i = -2i\lambda(dt)^{(\frac{2}{D_F})-1} \partial^i \ln \Psi. \tag{6}$$

Then, substituting Equation (6) in Equation (5), the geodesics equation (Equation (5)) (for details, see method from [42–44]) becomes a Schrödinger-type equation at various scale resolutions:

$$\lambda^2 (dt)^{(\frac{4}{D_F})-2} \partial^l \partial_l \Psi + i\lambda(dt)^{(\frac{2}{D_F})-1} \partial_t \Psi = 0. \tag{7}$$

The variable $\Phi = -2i\lambda(dt)^{(2/D_F)-1} \ln \Psi$ defines, through Equation (6), the complex scalar potential of the complex velocity field, while Ψ corresponds to the ablation plasma state of fractal type. Both variables, Φ and Ψ, have no direct physical meaning, but possible "combinations" of them can acquire it if they satisfy certain conservation laws.

Let us make explicit such a situation for Ψ. For this purpose, we first notice that the complex conjugate of Ψ, that is, $\overline{\Psi}$, satisfies through Equation (7) the following equation:

$$\lambda^2 (dt)^{(\frac{4}{D_F})-2} \partial^l \partial_l \overline{\Psi} - i\lambda(dt)^{(\frac{2}{D_F})-1} \partial_t \overline{\Psi} = 0. \tag{8}$$

Multiplying Equation (7) by $\overline{\Psi}$ and Equation (8) by Ψ, subtracting the results, and introducing the notations,

$$\rho = \Psi\overline{\Psi}, \quad J = i\lambda(dt)^{(2/D_F)-1}\left(\Psi \nabla \overline{\Psi} - \overline{\Psi} \nabla \Psi \right), \tag{9}$$

we can obtain the conservation law of the states density of fractal type:

$$\partial_t \rho + \nabla J = 0. \tag{10}$$

In Equation (10), ρ corresponds to the states density of fractal type and J corresponds to the states density current of fractal type.

4.3. Ablation Plasma Behavior through a Special Tunneling Effect of Fractal Type

In the following, we will perform an application of the mathematical model previously mentioned by analyzing one-dimensional (1-D) stationary dynamics in physical systems with spontaneous symmetry breaking in the form of a tunneling effect of fractal type. The results will be corelated/calibrated with the dynamics of an LPP from the perspective of generating two plasma structures, as well from the perspective of some essential characteristics of these structures.

Let us consider the following equation of fractal type [32–34]:

$$\lambda^2 (dt)^{(\frac{4}{D_F})-2} \partial^l \partial_l \Psi + i\lambda (dt)^{(\frac{2}{D_F})-1} \partial_t \Psi - \frac{U}{2}\Psi = 0, \tag{11}$$

which has an external restriction in the form of the scalar potential U. For the moment, the scalar potential is not explicitly defined.

In the 1-D stationary case, Equation (11) becomes

$$\lambda^2 (dt)^{(\frac{4}{D_F})-2} \partial_{zz}\Psi(z,t) + i\lambda (dt)^{(\frac{2}{D_F})-1} \partial_t \Psi(z,t) - \frac{U}{2}\Psi(z,t) = 0. \tag{12}$$

If the scalar potential U is time-independent, $\partial_t U = 0$, Equation (12) admits the stationary solution (for details on the method, see [42–44]):

$$\psi(z,t) = \theta(z) \exp\left[-\frac{i}{2m_0 \lambda (dt)^{(\frac{2}{D_F})-1}} Et\right], \tag{13}$$

where E is the fractal energy of the plasma entity, $\theta(x)$ is the stationary state of fractal type of the plasma entity, and m_0 is the fractal rest mass of the plasma entity. Then, $\theta(x)$ becomes a solution of the fractal non-temporal equation:

$$\partial_{zz}\theta(z) + \frac{1}{2m_0 \lambda^2 (dt)^{(\frac{4}{D_F})-2}} (E-U)\theta(z) = 0. \tag{14}$$

In the following, we will assume that the behavior of the LPP, such as self-structuring in two different substructures (Coulomb and thermal) based on the plasma plume separation on different oscillations modes, can be mimicked by spontaneous symmetry breaking. The mathematical procedure will imply the use of Equation (14) for a scalar potential U characteristic to the spontaneous symmetry breaking (see Figure 7). From the theory of spontaneous symmetry-breaking systems [31,42], the shape of the effective potential is already well-known and has the following expression:

$$V_{ef}(z) = V_0 \left[\left(\frac{z}{z_0}\right) - 1\right]^2, \quad V_0 = \text{const}, \quad z_0 = \text{const}.$$

The exact integral of the stationary Schrödinger equation of fractal type,

$$\partial_{zz}\theta(z) + \frac{1}{2m_0 \lambda^2 (dt)^{(\frac{4}{D_F})-2}}\left(E - V_0\left[\left(\frac{z}{z_0}\right)-1\right]^2\right) = 0,$$

is not achievable. The determination of the eigenfunctions and eigenvalues $\{\theta, E\}$ can be attained only by successive approximations, order by order, in the framework of stationary perturbation theory, starting from the 0 order, from the following eigenfunctions and eigenvalues:

$$\left\{\varphi_n(z), \; E_n = 2m_0\lambda(dt)^{(2/D_F)-1}\left(n+\frac{1}{2}\right)\right\}_{n\in N},$$

of the fractal harmonic linear oscillator. To avoid this setback, a simplifying hypothesis can be introduced considering some approximations, as the equation cannot be integrated in a closed form. The following statements can be made:

(i) At $|z| = 2|z_0|$, V_{ef} is much larger than V_0, which means to the right of z_0, and to the left of $-z_0$, the increase in the effective potential is fast enough so that it can be approximated with two infinite vertical walls at $z \pm l$, with $l \geq z_0$ (mandatory);

(ii) Because E_0 is placed below half of the local maxima V_0 and, at $|z| = 0.5|z_0|$, V_{ef} is already higher than $V_0/2$, the same thought process can done in $z \pm d$, with $d < z_0$ (mandatory), by placing in each of these points a vertical wall of V_0 height. Therefore, $V(\infty)$ for $U \equiv V(z)$ becomes a condition for the spontaneous symmetry breaking, as well as for U's symmetry.

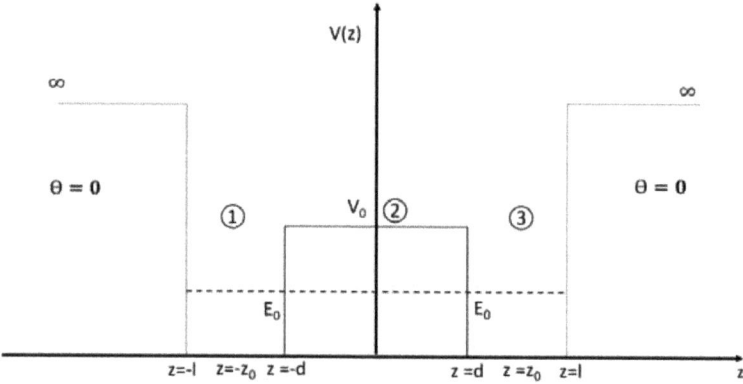

Figure 7. The effective potential in the case of the tunneling effect of fractal type for a physical system with spontaneous symmetry breaking (where E_0 is the fundamental energy level, V_0 is the height of the potential barrier, $z = 2d$ is the width of the potential barrier, and $z = l - d$ is the width of the potential well).

In these conditions, Equation (14) takes the following form:

$$\frac{d^2\theta_\alpha}{dz^2} + \frac{1}{2m_0\lambda^2(dt)^{(\frac{4}{D_F})-2}}[E - V_\alpha]\theta_\alpha = 0, \ \alpha = \overline{1,3}. \tag{15}$$

For each of the three regions, the solutions of Equation (15) are

$$\begin{aligned}\theta_1(z) &= C_+e^{ikz} + C_-e^{-ikz},\\ \theta_2(z) &= Be^{qz} + Ce^{-qz},\\ \theta_3(z) &= D_+e^{ikz} + D_-e^{-ikz},\end{aligned} \tag{16}$$

with

$$k = \left[\frac{E}{2m_0\lambda^2(dt)^{(\frac{4}{D_F})-2}}\right]^{\frac{1}{2}},$$

$$q = \left[\frac{V_0-E}{2m_0\lambda^2(dt)^{(\frac{4}{D_F})-2}}\right]^{\frac{1}{2}}, \tag{17}$$

and C_+, C_-, B, C, D_+, D_- are integration constants.

Due to the infinite potential in the two extreme regions, $|z| > l$, the state function of fractal type, $z = \pm l$, implies

$$\theta_2(-l) = C_+ e^{-ikl} + C_- e^{ikl} = 0,$$
$$\theta_3(l) = D_+ e^{ikl} + D_- e^{-ikl} = 0. \qquad (18)$$

As the states density of fractal type, $|\psi|^2$, is not altered by the multiplication of the state function of fractal type in the form of a constant phase factor of fractal type, the two equations for $C_\pm$ and $D_\pm$ can be immediately solved by imposing the forms,

$$C_+ = \tfrac{A}{2i} e^{ikl},\ C_- = -\tfrac{A}{2i} e^{-ikl},$$
$$D_+ = \tfrac{D}{2i} e^{-ikl},\ D_- = -\tfrac{D}{2i} e^{ikl}, \qquad (19)$$

so that $\theta_{1,3}$ are given through simple expressions:

$$\theta_1(z) = A\sin[k(z+l)],$$
$$\theta_3(z) = D\sin[k(z-l)]. \qquad (20)$$

These, along with ψ_2, lead to the concrete form of "alignment conditions" in $z = \pm d$:

$$\theta_1(-d) = \theta_2(-d),\ \theta_2(d) = \theta_3(d),$$
$$\tfrac{d\theta_1}{dz}(-d) = \tfrac{d\theta_2}{dz}(-d),\ \tfrac{d\theta_2}{dz}(d) = \tfrac{d\theta_3}{dz}(d), \qquad (21)$$

namely,

$$e^{-qd}B + e^{qd}C = A\sin[k(l-d)],$$
$$qe^{-qd}B - qe^{qd}C = kA\cos[k(l-d)] \text{ in } z = -d,$$
$$e^{qd}B + e^{-qd}C = -D\sin[k(l-d)],$$
$$qe^{qd}B - qe^{-qd}C = kD\cos[k(l-d)] \text{ in } z = d. \qquad (22)$$

Due to the algebraic form of the two equation pairs, in order to establish the concrete expression of the "secular equation" (for eigenvalues E of the energy), $\Delta[E] = 0$, we avoid calculating the 4th order determinant, $\Delta[k(E), q(E)]$, formed with the amplitude coefficients of fractal type, A, B, C, D, by employing the following: We note, with ρ, the ratio C/B, and we divide the first equation with the second one for each pair. It results in

$$\frac{e^{2qd}\rho+1}{e^{2qd}\rho-1} = -\tfrac{q}{k}tg[k(l-d)],$$
$$\frac{e^{-2qd}\rho+1}{e^{-2qd}\rho-1} = \tfrac{q}{k}tg[k(l-d)], \qquad (23)$$

which leads to the equation for ρ:

$$\frac{e^{2qd}\rho+1}{e^{2qd}\rho-1} + \frac{e^{-2qd}\rho+1}{e^{-2qd}\rho-1} = 0. \qquad (24)$$

We find

$$\rho^2 = 1,$$

which implies

$$\rho_- = -1,\ \rho_+ = 1. \qquad (25)$$

For $\rho_+ = 1$, the amplitude function of fractal type, $\theta_2(z) \cong \cosh(qz)$, is symmetric, similar to the states of fractal type with regard to the (spatial) reflectivity against the origin. Then, the permitted values equation of the fractal energy of these states, E_S, has the following concrete form:

$$\tan[k_S(l-d)] = -\frac{\coth(q_S d)}{q_S} k_S, \qquad (26)$$

where

$$k_S = \left[\frac{E_S}{2m_0\lambda^2(dt)^{(\frac{4}{D_F})-2}}\right]^{\frac{1}{2}},$$
$$q_S = \left[\frac{V_0-E_S}{2m_0\lambda^2(dt)^{(\frac{4}{D_F})-2}}\right]^{\frac{1}{2}}.$$
(27)

For $\rho_- = -1$, the amplitude function of fractal type $\theta_2(z) \cong \sinh(qz)$, so the states of fractal type will be antisymmetric, and the permitted values equation, E_A, becomes

$$\tan[k_A(l-d)] = -\frac{\tanh(q_A d)}{q_A}k_A,$$
(28)

where

$$k_A = \left[\frac{E_A}{2m_0\lambda^2(dt)^{(\frac{4}{D_F})-2}}\right]^{\frac{1}{2}},$$
$$q_A = \left[\frac{V_0-E_A}{2m_0\lambda^2(dt)^{(\frac{4}{D_F})-2}}\right]^{\frac{1}{2}}.$$
(29)

Now, some consequences are notable: The presence of the barrier (of finite height V_0) between $-d$ and d leads to the splitting of the fundamental level E_0 into two sublevels, E_S, E_A, accounting for the two states of fractal type, symmetric and antisymmetric, respectively, in which the system can be found. In the following, the above results will be calibrated to the LPP dynamics given by our experimental data. Therefore:

(i) At a global scale resolution, it can be seen as a self-structuring of the laser ablation plasma through its separation on two oscillation modes. We distinguish a Coulomb oscillations mode, corresponding to the fast structure, and a thermal oscillation mode, corresponding to the slow structure. The identification of the plasma structures, at a certain scale resolution, can be performed by assuming that the quantity $\Delta E = |E_A - E_S|$ is small compared to the plasma potential E_0, i.e., $\Delta E << E_0$, which implies the fact that $q_{A,S}$ are very close to

$$q_0 = \left[\frac{V_0-E_0}{2m_0\lambda^2(dt)^{(\frac{4}{D_F})-2}}\right]^{\frac{1}{2}},$$
(30)

and also considering that

$$\frac{\coth(q_0 d)}{\tanh(q_0 d)} = \left(\frac{e^{q_0 d}+1}{e^{q_0 d}-1}\right)^2 > 1,$$
(31)

with $d > 0$. This indicates that the fast structure is induced by the anti-symmetric energy state E_A, while the slower structure is induced by the symmetrical one E_S.

Potential barriers $V_0 - E_A$ and $V_0 - E_S$ are attributed to the double layers. In the empirical reality, the increase in the laser beam energy will lead to an increase in the intensity of the acceleration field during the Coulomb explosion [24] and, subsequently, of each of the two double layers; in the fractal geometry, this will be assimilated with the $V_0 - E_0$ parameters.

(ii) Assuming the functionality of the scale resolution superposition principle [32–34], at a global scale resolution (overall plasma plume), the following fractal equations are valid:

$$E_A = 2m_0\lambda(dt)^{(\frac{2}{D_F})-1}f_A,$$
$$E_S = 2m_0\lambda(dt)^{(\frac{2}{D_F})-1}f_S,$$
(32)

where f_A is the characteristic frequency of the Coulomb oscillation modes and f_S is the characteristic frequency of the thermal oscillation modes. As $E_A > E_S$, the oscillation frequency of the Coulomb mode will always be higher than that of the thermal oscillation mode, as it was shown experimentally in Figure 3.

(iii) At a global scale resolution, Equations (26) and (28) with the restrictions,

$$
\begin{aligned}
&k_A(l-d) \ll 1, \; q_A d \approx q_0 d \ll 1, \\
&k_S(l-d) \ll 1, \; q_S d \approx q_0 d \ll 1, \\
&l \gg d,
\end{aligned}
\tag{33}
$$

take the following approximate form:

$$
E_A \approx E_S \approx (V_0 - E_0) + \frac{6 m_0 \lambda^2 (dt)^{(\frac{4}{D_F})-2}}{l^2},
\tag{34}
$$

or even a simpler form, by multiplying with $1/2 m_0 \lambda (dt)^{(\frac{2}{D_F})-1}$:

$$
f_A \approx f_S = f \approx f_0 + \frac{3\lambda (dt)^{(\frac{2}{D_F})-1}}{l^2},
\tag{35}
$$

where the following notations have been used:

$$
\begin{aligned}
f &= \frac{E_A}{2 m_0 \lambda (dt)^{(\frac{2}{D_F})-1}} = \frac{E_S}{2 m_0 \lambda (dt)^{(\frac{2}{D_F})-1}}, \\
f_0 &= \frac{V_0 - E_0}{2 m_0 \lambda (dt)^{(\frac{2}{D_F})-1}}.
\end{aligned}
\tag{36}
$$

This leads to the value of f decreasing with the increase in l and increasing with the increase in the $(V_0 - E_0)$. In such a context, according to our previous considerations, the theoretical model at a local scale resolution $\lambda_A (dt)^{(\frac{2}{D_F})-1}$ and $\lambda_s (dt)^{(\frac{2}{D_F})-1}$ can describe the experimental behavior of the laser-produced plasmas presented in Figure 3.

(iv) In our model, the parameter $\lambda (dt)^{(2/D_F)-1}$ corresponds to the scale resolution. The parameter can be defined as:

$$
\lambda (dt)^{(2/D_F)-1} = \xi (dt)^{(2/D_F)-1} u(dt)^{(2/D_F)-1},
$$

where $\xi (dt)^{(2/D_F)-1}$ is the mean free path and $u(dt)^{(2/D_F)-1}$ is the expansion velocity of the laser ablation plasma. As the mean free path is inversely proportional to the collision cross-section, the collision cross-section has a functional dependence on the laser fluence. Therefore, through Equation (36), the oscillation frequency increases with the increase in the laser fluence.

(v) Considering that the energies from Equation (32) are of kinetic nature, $E_A = \frac{m_0 v_A^2}{2}$ and $E_S = \frac{m_0 v_S^2}{2}$, coupled with the fact that the separation of the plasma plume on the two oscillating modes can be expanded at local scale resolutions, $\lambda_A (dt)^{(\frac{2}{D_F})-1}$ and $\lambda_S (dt)^{(\frac{2}{D_F})-1}$, respectively, from Equation (32):

$$
f_A = \frac{v_A^2}{4 \lambda_A (dt)^{(\frac{2}{D_F})-1}}, \; f_S = \frac{v_S^2}{4 \lambda_S (dt)^{(\frac{2}{D_F})-1}},
\tag{37}
$$

where v_A is identified as the velocity of the fast structure and v_S is the velocity of the slower one. Particularly, for the ablation plasma dynamics associated with the scale transitions such as correlative-non-correlative processes (the fractal dimension of the movement curves has the value $D_F = 2$ [31]), Equation (37) becomes

$$
f_A = \frac{v_A^2}{4 \lambda_A}, \; f_S = \frac{v_S^2}{4 \lambda_S}.
\tag{38}
$$

Moreover,

$$\lambda_A = \frac{kT_e}{m\overline{v_A}}, \lambda_S = \frac{kT_e}{m\overline{v_S}}, \quad (39)$$

where k is the Boltzmann constant, T_e is the electron temperature, and m is the atomic mass of the ions. $\overline{v_A}$ is the ion–electron collision frequency at Coulomb-scale resolutions, and $\overline{v_S}$ is the collision frequency at thermal-scale resolutions, causing Equations (38) with (39) to become

$$f_A = \frac{m\overline{v_A}v_A^2}{kT_e}, f_S = \frac{m\overline{v_S}v_S^2}{kT_e}. \quad (40)$$

Let us now perform a numerical evaluation of Equation (40) using the experimental data as input parameters. The experimental data were presented in the previous section and the simulated results are synthesized in Table 1. We notice a good correlation between the experimental data and the simulated ones. Small discrepancies can be seen for the second structure at larger distances where the experimental data are noisier and the values can be affected by errors. In addition, this representation showcases once again the versatility of the fractal theoretical approach, as once the model is calibrated onto a specific measurement, it can offer results in very good agreement with the empirical data.

Table 1. Comparison between the experimental and theoretical values of the ionic oscillating frequency for the first (**a**) and second (**b**) plasma structure.

(a)	1st Structure Experimental Data (MHz)				1st Structure Simulated Data (MHz)			
Fluence (J/cm^2)	1 cm	2 cm	2.5 cm	3 cm	1 cm	2 cm	2.5 cm	3 cm
28	17.5 ± 0.2	15 ± 0.8	13 ± 0.5	7.6 ± 0.6	18.5 ± 0.3	16.1 ± 0.1	13.5 ± 0.2	9.4 ± 0.2
57	19.5 ± 0.3	16.8 ± 0.6	13.5 ± 0.7	8.5 ± 0.7	21.7 ± 0.2	18.2 ± 0.4	14.2 ± 0.6	10.6 ± 0.35
85	21 ± 0.7	19 ± 0.4	18 ± 0.5	16.3 ± 0.1	22.4 ± 0.05	20.2 ± 0.6	17.7 ± 0.5	16.3 ± 0.2
115	22 ± 0.1	19.26 ± 0.2	18.5 ± 0.3	17.3 ± 0.2	22.8 ± 0.5	20.9 ± 0.3	18.4 ± 0.2	17.4 ± 0.1
(b)	2nd Structure Experimental Data (MHz)				2nd Structure Simulated Data (MHz)			
Fluence (J/cm^2)	1 cm	2 cm	2.5 cm	3 cm	1 cm	2 cm	2.5 cm	3 cm
28	7.8 ± 0.1	6.5 ± 0.3	2 ± 0.4	1.2 ± 0.6	7.4 ± 0.6	6.44 ± 0.1	2.4 ± 0.3	2.2 ± 0.1
57	8.4 ± 0.2	7.2 ± 0.1	5.6 ± 0.1	4.58 ± 0.2	8.68 ± 0.6	7.28 ± 0.04	5.68 ± 0.05	4.35 ± 0.05
85	9.5 ± 0.4	9.3 ± 0.5	9 ± 0.3	8.2 ± 0.2	8.96 ± 0.6	9.08 ± 0.1	8.78 ± 0.04	8.14 ± 0.04
115	10 ± 0.5	9.8 ± 0.5	9.3 ± 0.4	8.6 ± 0.3	9.12 ± 0.6	9.36 ± 0.06	8.99 ± 0.05	8.4 ± 0.2

4.4. Mutual Conditionings of the Plasma Structures through Joint Invariant Functions

The fact that the presence of two plasma structures can be explained through the self-structuring of the plasma plume on two oscillation modes shows the mutual conditionings between these two structures. In order to characterize these conditionings, we must assume, according to the empirical proof presented in Figure 2, that the two plasma structures are characterized by two differential equations of dampened oscillator type for the charged particle densities. By using the group properties of these differential equations (invariances of SL(2R) type, algebras of Lie type, etc. [43]) and the Stoka theorem [41], we will build the joint invariant functions on the two isomorph SL(2R)-type groups. These functions will describe the mutual conditionings between the two plasma structures. Following this, let us consider the Lie algebra base associated with the SL(2R) group given by the infinitesimal generators [32,34]:

$$A_1 = \frac{\partial}{\partial h} + \frac{\partial}{\partial \overline{h}}, A_2 = h\frac{\partial}{\partial h} + \overline{h}\frac{\partial}{\partial \overline{h}}, A_3 = h^2\frac{\partial}{\partial h} + \overline{h}^2\frac{\partial}{\partial \overline{h}}, \quad (41)$$

and its commutation relations:

$$[A_1, A_2] = A_1, [A_2, A_3] = A_3, [A_3, A_1] = -2A_2, \quad (42)$$

where h is a complex amplitude and $\overline{h}$ is its complex conjugate. We assume that this group would characterize the Coulomb structure behavior.

Let us also consider the Lie algebra base associated with the SL(2R) group given by the infinitesimal generators:

$$B_1 = \frac{\partial}{\partial y} + \frac{\partial}{\partial \bar{y}}, \quad B_2 = y\frac{\partial}{\partial y} + \bar{y}\frac{\partial}{\partial \bar{y}}, \quad B_3 = y^2\frac{\partial}{\partial y} + \bar{y}^2\frac{\partial}{\partial \bar{y}}, \tag{43}$$

and its commutation relations:

$$[B_1, B_2] = B_1, \ [B_2, B_3] = B_3, \ [B_3, B_1] = -2B_2, \tag{44}$$

where y is a complex amplitude and $\bar{y}$ is its complex conjugate. We assume now that this group would characterize the thermal structure behavior.

The above-mentioned groups are isomorphic, i.e., the joint invariant functions F at the actions of these groups will be obtained as solutions of Stoka's equations [43]:

$$A_l F + B_l F = 0, \quad l = 1, 2, 3, \tag{45}$$

or, in a more explicit way,

$$\begin{array}{l}\frac{\partial F}{\partial h} + \frac{\partial F}{\partial \bar{h}} + \frac{\partial F}{\partial y} + \frac{\partial F}{\partial \bar{y}} = 0 \\ h\frac{\partial F}{\partial h} + \bar{h}\frac{\partial F}{\partial \bar{h}} + y\frac{\partial F}{\partial y} + \bar{y}\frac{\partial F}{\partial \bar{y}} = 0 \\ h^2\frac{\partial F}{\partial h} + \bar{h}^2\frac{\partial F}{\partial \bar{h}} + y^2\frac{\partial F}{\partial z} + \bar{y}^2\frac{\partial F}{\partial \bar{y}} = 0 \end{array} \tag{46}$$

The rank's system is 3; therefore, only one independent integral exists. This is the cross-ratio generated by means of the following relation:

$$\frac{h-y}{h-\bar{y}} : \frac{\bar{h}-y}{\bar{h}-\bar{y}} = \sigma, \tag{47}$$

where σ is real and positive. Now, any joint invariant function here is a regular function of this cross-section. In particular, for $\sigma = \tanh\tau$ with τ arbitrary, through Equation (47), z is connected to h by means of the following relation:

$$y = u + vh_0, \tag{48}$$

with

$$\begin{array}{l} h = u + iv \\ h_0 = -i\frac{\cosh\tau - e^{-i\alpha}\sinh\tau}{\cosh\tau + e^{-i\alpha}\sinh\tau} \\ \Delta\tau = 0 \end{array} \tag{49}$$

where Δ is the Laplace operator and α is real. Equations (48) and (49) establish a relation between the current density amplitude of the two plasma structures. Moreover, through Equation (49), we showcase a self-modulation in amplitude of the ionic current density through the Stoler-type transformation [44,45]. We showcase Stoler-type transformations with an important role in the theory of coherent states, meaning on the charge ionization/neutralization processes [43,44]. The possibility to use Stoler-type transformation for the analysis of laser-produced plasma dynamics has more profound implications. The existence of coherent states implies that the individual ejected particles are coherent (through the double-layer characteristic to each laser ablation plasma structure), and there is a connecting tissue amongst the ablated particle cloud. This aspect of the laser-produced plasma was empirically showcased by our group in [10], where we show that at all laser ablation scales, there are mathematical functions connecting series of materials, a property that is inherent to a fractal model. As the particularity of a fractal model is the use of zoom in/zoom out processes, we can still find a coherence between the plasma structures (Coulomb or thermal). This is seen experimentally as, although the individual values differ for each structure, the overall trends are respected regardless of the nature of the structure. Another important role is played by the ionization/neutralization processes. The ionization processes are directly connected to the collisions during expansion, accounted for in the framework of our fractal model through the fractalization degree and the dependences presented in Equation (36). Our mathematical approach is, therefore, a complex one accounting for the main processes occurring during the expansion of the laser-produced plasma, and, through an adequate calibration of the model, we can offer a quantitative analysis of the oscillatory phenomena.

5. Conclusions

Transient plasmas generated by ns-laser ablation on an Mg target at high irradiation fluences were investigated by means of the Langmuir probe method and fractal analysis. The electrical investigations revealed the presence of a dual structure for the ionic and electronic saturation currents. An oscillatory behavior was observed for short evolution times and a classical quasi-exponential decrease was observed for longer times. Two oscillation frequencies were found in all irradiation conditions: A few tens of MHz and a few MHz. Each frequency is associated with an individual plasma structure. Space- and time-resolved investigations were performed, revealing an exponential decrease in the main plasma parameters during expansion coupled with an increase in the plasma potential at high measurement distances. The effect of the laser fluence on both oscillatory and classical plasma parameters was investigated. Similar empirical evolutions were found to be described by a steep increase followed by saturation at higher fluences.

The dynamics of laser-produced plasmas were described through the fractal theory of motion given by Schrödinger regimes of fractal type. The calibration of such dynamics through a fractal-type tunneling effect for physical systems with spontaneous symmetry breaking allows the self-structuring of laser-produced plasma in the two structures (Coulomb and thermal substructures) based on its separation on different oscillations modes, as well as the determination of some characteristics involved in the self-structuring process: The evolution of the structure's oscillation frequencies with the target probe distance, and with the laser fluence, the average values of these frequencies. The mutual conditionings between the two structures were given as a joint invariant function on the action of two isomorphous groups of SL(2R). Their isomorphism implies structure amplitude self-modulation through Stoler-type transformation (i.e., through charge generation/neutralization processes).

Author Contributions: Conceptualization, I.M., S.A.I. and M.A.; methodology, I.M., S.A.I. and M.A.; investigation, I.M., S.A.I. and M.A.; writing—original draft preparation, I.M., S.A.I., F.N. and M.A.; writing—review and editing, I.M., S.A.I., F.N. and M.A.; visualization, S.A.I. and I.M.; supervision, M.A. All authors have read and agreed to the published version of the manuscript.

Funding: This work was financially supported by the National Authority for Scientific Research and Innovation in the framework of the Nucleus Program –16N/2019 and by a grant of the Ministry of Research and Innovation, CNCS-UEFISCDI, project number PN-III-P4-ID-PCE-2016-0355, within PNCDI III.

Conflicts of Interest: The authors declare no conflict of interest.

References

1. Dijkkamp, D.; Venkatesan, T.; Wu, X.D.; Shaheen, S.A.; Jisrawi, N.; Min-Lee, Y.H.; McLean, W.L.; Croft, M. Preparation of Y-Ba-Cu oxide superconductor thin films using pulsed laser evaporation from high T_c bulk material. *Appl. Phys. Lett.* **1987**, *51*, 619–621. [CrossRef]
2. Bulai, G.; Trandafir, V.; Irimiciuc, S.A.; Ursu, L.; Focsa, C.; Gurlui, S. Influence of rare earth addition in cobalt ferrite thin films obtained by pulsed laser deposition. *Ceram. Int.* **2019**, *45*, 20165–20171. [CrossRef]
3. Craciun, D.; Socol, G.; Stefan, N.; Dorcioman, G.; Hanna, M.; Taylor, C.R.; Lambers, E.; Craciun, V. Applied Surface Science The effect of deposition atmosphere on the chemical composition of TiN and ZrN thin films grown by pulsed laser deposition. *Appl. Surf. Sci.* **2014**, *302*, 124–128. [CrossRef]
4. Yang, Q.I.; Zhao, J.; Zhang, L.; Dolev, M.; Fried, A.D.; Marshall, A.F.; Risbud, S.H.; Kapitulnik, A. Pulsed laser deposition of high-quality thin films of the insulating ferromagnet EuS. *Appl. Phys. Lett.* **2014**, *104*, 082402. [CrossRef]
5. Craciun, V.; Doina, C. Reactive pulsed laser deposition of TiN. *Appl. Surf. Sci.* **1992**, *54*, 75–77. [CrossRef]

6. Dorcioman, G.; Socol, G.; Craciun, D.; Argibay, N.; Lambers, E.; Hanna, M.; Taylor, C.R.; Craciun, V. Applied Surface Science Wear tests of ZrC and ZrN thin films grown by pulsed laser deposition. *Appl. Surf. Sci.* **2014**, *306*, 33–36. [CrossRef]
7. Amoruso, S.; Bruzzese, R.; Velotta, R.; Spinelli, M.; Vitiello, M.; Wang, X. Characterization of $LaMnO_3$ laser ablation in oxygen by ion probe and opticalemission spectroscopy. *Appl. Surf. Sci.* **2005**, *248*, 45–49. [CrossRef]
8. Irimiciuc, S.; Bulai, G.; Agop, M.; Gurlui, S. Influence of laser-produced plasma parameters on the deposition process: In situ space- and time-resolved optical emission spectroscopy and fractal modeling approach. *Appl. Phys. A* **2018**, *124*, 1–14. [CrossRef]
9. Aragón, C.; Aguilera, J.A. Characterization of laser induced plasmas by optical emission spectroscopy: A review of experiments and methods. *Spectrochim. Acta Part B* **2008**, *63*, 893–916. [CrossRef]
10. Irimiciuc, S.A.; Nica, P.E.; Agop, M.; Focsa, C. Target properties–plasma dynamics relationship in laser ablation of metals: Common trends for fs, ps and ns irradiation regimes. *Appl. Surf. Sci.* **2019**, *506*, 144926. [CrossRef]
11. Irimiciuc, S.A.; Gurlui, S.; Bulai, G.; Nica, P.; Agop, M.; Focsa, C. Langmuir probe investigation of transient plasmas generated by femtosecond laser ablation of several metals: Influence of the target physical properties on the plume dynamics. *Appl. Surf. Sci.* **2017**, *417*, 108–118. [CrossRef]
12. Irimiciuc, S.A.; Mihaila, I.; Agop, M. Experimental and theoretical aspects of a laser produced plasma. *Phys. Plasmas* **2014**, *21*, 093509. [CrossRef]
13. Chen, J.; Lippert, T.; Ojeda-G-P, A.; Stender, D.; Schneider, C.W.; Wokaun, A. Langmuir probe measurements and mass spectrometry of plasma plumes generated by laser ablation of $La_{0.4}Ca_{0.6}MnO_3$. *J. Appl. Phys.* **2014**, *116*, 073303.
14. Kumari, S.; Kushwaha, A.; Khare, A. Spatial distribution of electron temperature and ion density in laser induced ruby (Al_2O_3:Cr^{3+}) plasma using Langmuir probe. *J. Instrum.* **2012**, *7*, C05017. [CrossRef]
15. Schou, J.; Toftmann, B.; Amoruso, S. Dynamics of a laser-produced silver plume in an oxygen background gas. In *High-Power Laser Ablation V*; International Society for Optics and Photonics: New Mexico, NM, USA, 2004; Volume 5448, pp. 22–26.
16. Harilal, S.S.; Bindhu, C.V.; Tillack, M.S.; Najmabadi, F.; Gaeris, A.C. Internal structure and expansion dynamics of laser ablation plumes into ambient gases. *J. Appl. Phys.* **2003**, *93*, 2380–2388. [CrossRef]
17. Harilal, S.S.; Issac, R.C.; Bindhu, C.V.; Nampoori, V.P.N.; Vallabhan, C.P.G. Temporal and spatial evolution of C_2 in laser induced plasma from graphite target. *J. Appl. Phys.* **1996**, *80*, 3561. [CrossRef]
18. Vivien, C.; Hermann, J.; Perrone, A.; Boulmer-Leborgne, C. A study of molecule formation during laser ablation of graphite in low-pressure ammonia. *J. Phys. D* **1999**, *32*, 518–528. [CrossRef]
19. Thestrup, B.; Toftmann, B.; Schou, J.; Doggett, B.; Lunney, J.G. Ion dynamics in laser ablation plumes from selected metals at 355 nm. *Appl. Surf. Sci.* **2002**, *197*, 175–180. [CrossRef]
20. Harilal, S.S.; Bindhu, C.V.; Nampoor, V.P.N.; Vallabhan, C.P.G. Temporal and spatial Behavior of electron density and temperature in a laser-produced plasma from $YBa_2Cu_3O_7$. *Appl. Spectrosc.* **1998**, *52*, 449–455. [CrossRef]
21. Nica, P.; Agop, M.; Gurlui, S.; Focsa, C. Oscillatory Langmuir probe ion current in laser-produced plasma expansion. *EPL* **2010**, *89*, 65001. [CrossRef]
22. Tang, E.; Xiang, S.; Yang, M.; Li, L. Sweep Langmuir Probe and Triple Probe Diagnostics for Transient Plasma Produced by Hypervelocity Impact. *Plasma Sci. Technol.* **2012**, *14*, 747–753. [CrossRef]
23. Singh, S.C.; Fallon, C.; Hayden, P.; Mujawar, M.; Yeates, P.; Costello, J.T. Ion flux enhancements and oscillations in spatially confined laser produced aluminum plasmas. *Phys. Plasmas* **2014**, *21*, 093113. [CrossRef]
24. Bulgakov, A.V.; Bulgakova, N.M. Dynamics of laser-induced plume expansion into an ambient gas during film deposition. *J. Phys. D* **1999**, *28*, 1710–1718. [CrossRef]
25. Focsa, C.; Gurlui, S.; Nica, P.; Agop, M.; Ziskind, M. Plume splitting and oscillatory behavior in transient plasmas generated by high-fluence laser ablation in vacuum. *Appl. Surf. Sci.* **2017**, *424*, 299–309. [CrossRef]
26. Irimiciuc, S.A.; Agop, M.; Nica, P.; Gurlui, S.; Mihaileanu, D.; Toma, S.; Focsa, C. Dispersive effects in laser ablation plasmas. *Jpn. J. Appl. Phys.* **2014**, *53*, 116202. [CrossRef]
27. Irimiciuc, Ş.; Enescu, F.; Agop, A.; Agop, M. Lorenz Type Behaviors in the Dynamics of Laser Produced Plasma. *Symmetry* **2019**, *11*, 1135. [CrossRef]

28. Irimiciuc, S.A.; Gurlui, S.; Nica, P.; Focsa, C.; Agop, M. A compact non-differential approach for modeling laser ablation plasma dynamics. *J. Appl. Phys.* **2017**, *12*, 083301. [CrossRef]
29. Marine, W.; Bulgakova, N.M.N.M.; Patrone, L.; Ozerov, I. Electronic mechanism of ion expulsion under UV nanosecond laser excitation of silicon: Experiment and modeling. *Appl. Phys. A* **2004**, *79*, 771–774. [CrossRef]
30. Merlino, R.L. Understanding Langmuir probe current-voltage characteristics. *Am. J. Phys.* **2007**, *75*, 1078. [CrossRef]
31. Nottale, L. *Scale Relativity and Fractal Space-Time: An Approach to Unifying Relativity and Quantum Mechanics*; Imperial College Press: London, UK, 2011.
32. Merches, I.; Agop, M. *Differentiability and Fractality in Dynamics of Physical Systems*; World Scientific: Singapore, 2016.
33. Agop, M.; Paun, V.P. *On the New Paradigm of Fractal Theory*; Fundamental and applications, Romanian Academy Publishing House: Bucharest, Romania, 2017.
34. Agop, M.; Merches, I. *Operational Procedures Describing Physical Systems*; CRC Press, Taylor and Francis Group: London, UK, 2019.
35. El-Nabulsi, R.A. Some implications of position dependent mass quantum fractional Hamiltonian in Quantum mechanics. *Eur. Phys. J. Plus* **2019**, *134*, 192. [CrossRef]
36. Wójcik, D.; Białynicki-Birula, I.; Życzkowski, K. Time Evolution of Quantum Fractals. *Phys. Rev. Lett.* **2000**, *85*, 5022. [CrossRef]
37. Laskin, N. Fractals and quantum mechanics. *Chaos* **2000**, *4*, 780–790. [CrossRef]
38. Golmankhaneh, A.K.; Golmankhaneh, A.K.; Balean, D. About Schrödinger Equation on Fractals CurvesImbedding inR^3. *Int. J. Theor. Phys.* **2015**, *54*, 1275–1282. [CrossRef]
39. Chuprikov, N.L.; Spiridonov, O.V. A new type of solution of the Schrödinger equationon a self-similar fractal potential. *J. Phys. A* **2008**, *41*, 409801. [CrossRef]
40. Dhillon, H.S.; Kusmartsev, F.V.; Kürten, K.E. Fractal and Chaotic Solutions of the Discrete Nonlinear Schrödinger Equation in Classical and Quantum Systems. *J. Nonlinear Math. Phys.* **2001**, *8*, 38–49. [CrossRef]
41. El-Nabulsi, R.A. Path Integral Formulation of Fractionally Perturbed Lagrangian Oscillators on Fractal. *J. Stat. Phys.* **2018**, *172*, 1617–1640. [CrossRef]
42. Mandelbrot, B. *The Fractal Geometry of Nature*; WH Freeman Publisher: New York, NY, USA, 1993.
43. Stoka, M.I. *Integral Geometry*; Romanian Academy Publishing House: Bucharest, Romania, 1967.
44. Stoler, D. Equivalence Classes of Minimum Uncertainty Packets. *Phys. Rev. D* **1970**, *1*, 3217. [CrossRef]
45. Stoler, D. Equivalence Classes of Minimum-Uncertainty Packets. II. *Phys. Rev. D* **1972**, *4*, 1925. [CrossRef]

© 2020 by the authors. Licensee MDPI, Basel, Switzerland. This article is an open access article distributed under the terms and conditions of the Creative Commons Attribution (CC BY) license (http://creativecommons.org/licenses/by/4.0/).

Article

Investigations of Laser Produced Plasmas Generated by Laser Ablation on Geomaterials. Experimental and Theoretical Aspects

Florin Enescu [1], Stefan Andrei Irimiciuc [2,*], Nicanor Cimpoesu [3], Horea Bedelean [4], Georgiana Bulai [5], Silviu Gurlui [1] and Maricel Agop [6,7]

1. Faculty of Physics, LOA-SL, Alexandru Ioan Cuza University of Iasi, Iasi 700506, Romania; florinenescu@yahoo.com (F.E.); sgurlui@uaic.ro (S.G.)
2. Institute for Laser, Plasma and Radiation Physics, Bucharest 077125, Romania
3. Faculty of Materials Science and Engineering, "Gh. Asachi" Technical University from Iasi, Iasi 700050, Romania; nicanorcimpoesu@gmail.com
4. Department of Geology, Babes-Bolyai University, Cluj-Napoca 400084, Romania; hbedelean@gmail.com
5. Integrated Center for Studies in Environmental Science for North-East Region (CERNESIM), Alexandru Ioan Cuza University of Iasi, Iasi 700506, Romania; georgiana.bulai@uaic.ro
6. Department of Physics, "Gh. Asachi" Technical University of Iasi, Iasi 700050, Romania; m.agop@yahoo.com
7. Romanian Scientists Academy, Bucharest 050094, Romania
* Correspondence: stefan.irimiciuc@inflpr.ro

Received: 21 October 2019; Accepted: 7 November 2019; Published: 9 November 2019

Abstract: Several surface investigation techniques, such as X-ray diffraction (XRD), EDX, and optical microscopy, were employed in order to describe the mineral contents in several geomaterials. Space and time resolved optical emission spectroscopy was implemented to analyze the plasma generated by the laser–geomaterial interaction. The values of the plasma parameters (velocity and temperature) were discussed with respect to the nature of the minerals composing the geomaterials and the morphological structure of the samples. Correlations were found between the excitation temperatures of the atomic and ionic species of the plasmas and the presence of calcite in the samples. A mathematical model was built to describe the dynamics in ablation plasma using various mathematical operational procedures: multi structuring of the ablation plasma by means of the fractal analysis and synchronizations of the ablation plasma entities through SL (2R) type group invariance and in a particular case, through self-modulation in the form of Stoler type transformations. Since Stoler type transformations are implied in general, in the charge creation and annihilation processes, then the SL (2R) type group invariance become fundamental in the description of ablation plasma dynamics.

Keywords: optical emission spectroscopy; laser ablation; petrographic analysis; fractal model; group invariance

1. Introduction

One of the new emerging applications of laser ablation is laser-induced breakdown spectroscopy (LIBS) with implementation in environmental science, space applications or food industry [1]. LIBS can analyze a wide range of samples spanning from metals, semiconductors, glasses, biological tissues, plastics, soils, and plants, to thin layer paint coatings or electronic materials [2]. From the perspective of analyzing geo-materials [3] this technique has its advantages: Fast to no preparation of samples which also means no special sample preparation skills are required, potential for in situ analysis, small sample size requirements, the samples can be in a solid [3], liquid or gas state [4], and sensitive to light elements as H, Be, Li, B, C, N, O, Na, and Mg [5]. Successful applications of the LIBS technique to the analysis of range of metallic targets (aluminum alloys, iron-based alloys, copper-based alloy, precious alloys) and a summary of good practices were discussed in the works of Palleski et al. [6]. The performance of such a technique is strongly dependent on the complex processes involved during plasma formation and expansion, as the study object for LIBS is the strong emitting laser induced plasmas. As such differential absorption by the material and by the particle vapor, or the matrix effect [7] can strongly affect the properties of the resultant plasma (usually confined to a few mm from the sample and expanding with high velocities). The matrix effect issues are shown to be overcome by different sample preparation techniques (i.e., in the form of pressed pelletized powder [8] or, with much better results, fused glass), however these are breaking on the main advantage of LIBS—reduced time consuming sample preparation—which is lost and the operational costs are higher due to the special equipment needed. The implementation of LIBS technique on soils needs also to consider the influences of compression force, moisture, and total content of easily ionized elements on line intensities and electron density were studied and because dramatic changes occur, the recommendations were to use dry samples as much as possible and observe plasma as early it can be done in order to minimize the matrix effects on results [9].

In case of qualitative elemental analysis and to simultaneous deal with the complexity of data, statistical methods like partial least squares discriminant analysis [10,11] (PLSDA) are used to obtain information about unknown samples.

The practicality of the technique is often shown in the little preparation of the diagnostics apparatus itself with approaches like calibration free LIBS (CF-LIBS) [1] which manages to produce closer values than Monte Carlo simulated annealing optimization method (MC-LIBS) [12] with respect to a standard measurement. Aside their many advantages there are still some important drawbacks to this approach, mainly the need for accurate plasma parameters of the investigated area. Real experimental conditions have proven a deviation from ideal conditions (optically thin, in LTE and spatio–temporal homogenous). Therefore, errors are introduced by experimental aberrations and inaccuracy of spectral data, and it was found that they are contributing to the overall uncertainty on the quantitative results more than theoretical parameters (inaccuracy of measurements of detector spectral efficiency weighs more on the results than a typical uncertainty in the electron density value) [13]. As the LIBS shifts to non-ideal plasmas, one major issue is self-absorption in thick plasmas in CF-LIBS, this phenomenon was investigated, and correction procedures were developed to ensure the reliability of results. A recursive algorithm was created to consider the non-linear self-absorption effects occurring in the plasma and this extended the range of application of the CF-LIBS method [14].

For CF-LIBS the presence of local thermodynamic equilibrium (LTE) is imperative as such estimation of electron density and plasma temperature, have a huge impact on the overall result, especially with strong heterogeneity in the ratio of the emitted lines. A special attention needs to be given to Boltzmann plot method implemented for each composing element, which for complex minerals and geomaterials can be an arduous job, thus the need of combining different types of investigations for both accurate plasma parameter determination and elemental identification. Data fusions of LIBS results with complementary methods of analysis like Raman and reflectance spectroscopy, X-Ray fluorescence were performed on iron ore, and remote sensing instrument suite was integrated in the Mars 2020 Rover [11]. There are other approaches like LIBS as quantitative analysis for materials located relatively far away from the LASER source. This requirement comes mandatory when dealing with potentially dangerous or out of reach radioactive or extraterrestrial materials. Mineral discrimination can also be performed with the help of PLSDA statistical method [11].

Another domain of actual and acute interest is the detection of explosive residues using LIBS in standoff mode of operation. Organic and inorganic explosive residues placed up to 30 meters behind transparent barriers (polymethylmethacrylate and glasses) were successfully detected without false positives with 8 shots as long as the laser beam energy can go through the barrier and part of the plasma light can be collected [15].

In this work we use a combinatorial plasma diagnosis approach in order to expand the quantity of information provided by LIBS. The approaches are used for mineral identification purposes, with the aim here, being the investigations of the relationships between plasma parameters in terms of temperature and velocity, and the nature of minerals. For this purpose, four geomaterials previously collected from various locations within the Northern Hemisphere were analyzed. For all minerals we implemented step-by-step investigation methods that expanded from polarized light microscopy, EDX spectroscopy and X-Ray Diffraction Spectroscopy. The techniques allowed the identification of the composing minerals and offered information about the heterogeneity of the investigated samples. Finally, the samples were irradiated with high power ns beams, and the light emitted from the laser induced plasma was investigated through space-and time-resolved investigation techniques. The complementary investigation methods used here, aid us in founding a strong possible correlation between parameters like excitation temperature or expansion velocity with the abundance of calcite structure. Also, a mathematical model to describe the ablation plasma dynamics using various mathematical operational procedures (fractal analysis, group invariance, differentiable geometry in Lobacewski, etc.) is built. From such a perspective the SL (2R) type group invariance can become fundamental in describing complex phenomena in laser ablation plasmas.

2. Materials and Methods

2.1. Samples Details

Coordinates of the places from where the samples were collected are given below along with the potential expectations in terms of composition, structure, and formation, information taken from the literature review. This cannot reflect or aspire to fully describe our particular samples but will provide a sense of perspective to each subsequent analysis. As most of the investigated geomaterials are found to present a wide range of minerals which could characterize an entire geographical area it becomes important not only to know the structure of our samples but also to get information on an entire family of rocks that could possibly be found in the vicinity of our collected samples.

Sample #1—the collected sample was taken from Petra, Jordan [16], Coordinates: 30°20′11.8″ N 35°25′59.3″ E. Petra area is located in the Western Jordan, in a large rift valley which extend from Gulf of Aqaba, northwards along the Wadi Araba to the Dead Sea and the Jordan Valley. The sample from Petra is assigned to the Umm Ishrin Formation, part of a larger sequence called the Ram Group of Cambrian age. The sediments are deposited on the igneous, crystalline basement rock formed during the Precambrian that are exposed in the mountains outside the rift valley. Red Cambrian sandstones are about 300 m thick [16]. The Umm Ishrin Formation consists of medium-to-coarse-grained, well to moderately sorted, poorly cemented quartz arenites, with minor amounts of siltstone and mudstone [17,18]. The color is generally reddish brown, with hues of red, orange, yellow and white, due to the presence of iron oxides, mainly limonite. The mineralogical composition of red sandstones consists mainly of quartz (up to 95%), and then, in small amounts, feldspar and micas [18,19]. It is considered to be of fluvial origin [19].

Sample #2—the collected sample was from Bowen Island, British Columbia, Canada (42°46′50.9″ N 0°27′22.7″ W). Bowen Island is located on the west flank of the Coast range batholith of British Columbia, at the junction of Howe Sound and the Gulf of Georgia. The rocks from Bowen Island are of igneous origin, of an extrusive and intrusive nature. In the main, the rocks consist of a volcanic assemblage of great thickness, made up of flows, breccias, agglomerates and tuffaceous sediments cut by basic porphyry dykes [20]. Most of the rocks from the island belong to the Bowen Island Group of Lower Jurassic age [21], consisting of mafic to intermediate volcanic rocks (basalts and andesite) formed as lavas, shallow intrusions, volcanic ash deposits, interbedded with volcanoclastic sandstone, siliceous argillite, chert and tuff, and intruded by sills and shallow level intrusions. Because they are resistant to erosion, they can form prominent hills. The volcanic formation is deformed and form tight east trending folds, being intruded by granodiorite and monzonite of Middle and Late Jurassic age [22]. On the eastern part of Bowen Island, where the sample was collected, massive meta-basalt flows and sills, and andesitic feldspar porphyry intrusions are found [22]. On Bowen Island, dark green, fine-grained andesites are composed of albitized plagioclase and hornblende, variably altered to opaque minerals, epidote, and calcite. These rocks are locally interbedded with thinly laminated to massive fine-grained siliceous tuff [17]. The volcanic sequence and Jurassic structures are cut by a north-trending quartz feldspar porphyry stock of rhyodacite composition.

Sample #3—the collected sample was taken from a reddish formation near Pico Anayet [18], 2575 m, Pyrenees, Spain (42°46′50.9″ N 0°27′22.7″ W). The Anayet Massif is an E–W axis mountain range located in the central segment of the Pyrenean axial zone (northern Spain), displaying an extensive sedimentary record of Stephanian–Permian deposits. This large outcrop of late-hercynian materials is surrounded by Devonian slates, Lower Carboniferous greywackes, and several types of pyroclastic rocks (deformed during the Hercynian orogeny). The Permian deposits are represented by several thousand meter-thick series of continental sediments, volcanoclastic and volcanic sediments [23]. The continental detrital deposits between Stephanian and Lower Triassic have been described as post-Hercynian molasses [23]. These deposits have been classically divided in four main detrital groups, mainly composed of arenites (sandstones), conglomerates and lutites, with three basic volcanic episodes interbedded [24].

Sample #4—the collected sample was taken from the top of Pico Collarada, 2886 m, Pyrenees, Spain (42°42′51.88″ N, 0°28′14.81″ W). The analyzed sample was collected from the Pyrenees Mountains area, about 17 km north of Jaca (Spain). The Pyrenean orogenic belt resulted from the collision between the Iberian and the European plates from Late Cretaceous to Miocene times. The Pyrenees consists of an Axial Zone, composed of Paleozoic rocks—granitoids and metamorphic rocks, bounded by the North Pyrenean and South Pyrenean zones, where Mesozoic and Cenozoic rocks—clastic and carbonate rocks, are found [25]. The Paleogene sequence from the southern side of the Pyrenean orogen consists of Paleocene to Eocene light-coloured massive limestones, turbidites (Lower–Middle Eocene flysch) and coastal, non-marine deposits [26] (Upper Eocene–Lower Oligocene molasse). The turbidite systems (Lutetian) are built up by a succession of sandstones and mudstones including carbonate megaturbidites [27].

2.2. Characterization Methods

X-ray diffraction (XRD) analyses were performed using a Bruker D8 Advance diffractometer equipped with a Cu Kα anticathode. The XRD spectra were recorded from 5° to 65° 2θ degree (40 kV, 40 mA). The ICDD Database PDF-2/Release 2012 was used to specify the values of the reflection peaks. Optical microscopy analyses were performed on a Zeiss AxioLab microscope. A scanning electron microscope model Vega LMH II–Tescan®, coupled with an EDX detector, model Quantax QX2-Bruker-Roentec® allowed the recording of the micrographic analyzed area and the spectrum in order to investigate the structure distribution and elemental composition. The EDX analyses were performed on an area of 1 mm^2 for each sample. For an individual analysis and the determination of the standard deviation, 10 experiments were performed on each area of the experimental samples.

Optical emission spectroscopy analysis was performed using the experimental set up described in [28,29]. The second harmonic with λ = 532 nm of a 10 ns Nd: YAG pulsed laser beam (Brilliant EaZy) was focused by a f = 40 cm lens on the samples. The spot diameter at the impact point was around 700 μm while the laser fluence was 19 J/cm^2. The samples were placed in a vacuum chamber where the pressure (5·10^{-2} Torr) was maintained using a 300 L/m dry scroll pump (Agilent TriScroll 300). The formation and dynamics of the plasma plume were studied by an intensified ICCD camera: PI-MAX3, 1024i with a gate time of 30 ns, placed orthogonal to the plasma expansion direction, coupled to a Princeton Instruments Acton 2750 monochromator (with a resolution of 0.2 nm). Two experimental approaches were considered: In the first one the overall emission of the plasma was studied by ICCD fast camera imaging (each image was averaged over 20 events) while in the second one the spectrally resolved emission was recorded (with each spectrum being averaged over 1000 events). For the later, a step-and-glue procedure was used to record the global emission spectra (gate width of 2 μs and a gate delay of 100 ns) in a 300–700 nm spectral range from a 600 μm wide plasma volume centered on the main expansion direction. The recording of each spectrum was preceded by the collection of the background noise and the subsequent subtraction of it from the data. During the experiments the targets were continuously moved to ensure fresh surfaces and to overcome the heterogeneity of the investigated rocks.

3. Results and Discussions

3.1. Structural Investigations

For the identification of individual minerals found in the four samples, a series of complementary surface techniques was considered. Optical microscopy in polarized light is a fast analyzing technique which can be used to identify the individual minerals of the rocks and offers a general idea about the heterogeneity of the surface which is one of the main factors that has to be considered before a technique like LIBS is implemented. The results of these investigations are presented in Figure 1a–d.

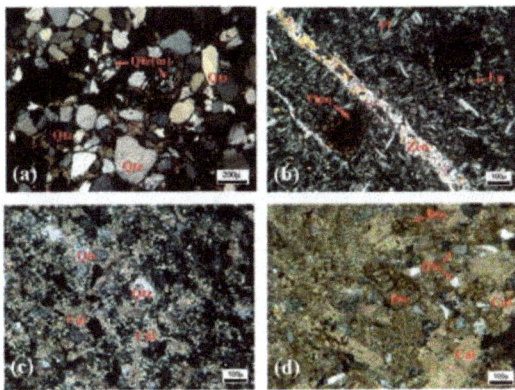

Figure 1. Optical microscopy and polarized microscopy images of all the investigated samples. Bio—bioclasts; Cal—calcite; Ep—epidote; Opq—opaque mineral; Pl—plagioclase; Qtz—quartz; Qtz(m)—metamorphic quartz; Zeo—zeolites. ((**a**) Sample #1—Quartz sandstone, (**b**) Sample #2—Spilite, (**c**) Sample #3—Sandstone, (**d**) Sample #4—Extraclastic bioclastic limestone).

Sample #1 (Figure 1a) is identified as a sedimentary rock [16]—quartz sandstone—that contains sand-sized grains (0.063–2 mm). The quartz grains (SiO_2) can be composed of single crystal or can be polycrystalline (fragments of quartzite—a metamorphic rock). All grains are moderately rounded. The carbonate cement (calcite—$CaCO_3$) fills the spaces between the quartz granules. Some limonite ($FeO(OH) \cdot nH_2O$) staining can be observed, which causes the reddish color of the rock, the full picture is completed with some traces of Kaolinite ($Al_2Si_2O_5(OH)_4$) observed only on XRD (Figure 2a).

Sample #2 (Figure 1b) is a basalt (spilite), an extrusive igneous (volcanic) rock. Spilite is an igneous rock produced when basaltic lava reacts with seawater, or is formed by hydrothermal alteration when sea water circulates through hot volcanic rocks. Under the microscope, the sample rock is composed of phenocrysts of plagioclase and very rare relict pyroxenes in a fine-grained holocrystalline groundmass made up of plagioclases, epidote ($Ca_2(Fe^{3+}, Al)_3(SiO_4)_3(OH)$), opaque minerals, and iron oxides/hydroxides. Different types of alteration can be observed: primary feldspar has been transformed into albite ($NaAlSi_3O_8$); pyroxenes and plagioclases have been replaced by other minerals such as epidote/zoisite (in thin sections, some "nests" of epidote/zoisite surrounded by a microlitic mass consists of the same minerals could be observed); secondary minerals as the result of alteration could be commonly chlorite ($Mg_5Al(AlSi_3O_{10})(OH)_8$) and iron oxides/hydroxides (hematite, limonite).

Cavities filled with secondary minerals (probably zeolites, phillipsite—[(K,Na,Ca)$_{1-2}$(Si,Al)$_8$O$_{16}$·6H$_2$O]) could be observed in thin sections. Actinolite (Ca$_2$(Mg,Fe)$_5$Si$_8$O$_{22}$(OH)$_2$) (observed only on XRD, Figure 2b) is a secondary mineral as a fine-grained alteration product of pyroxene.

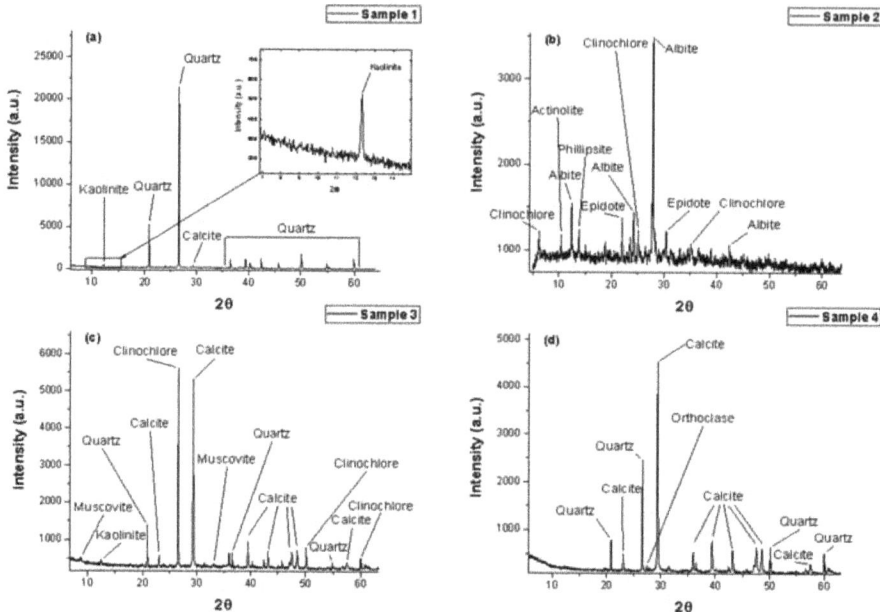

Figure 2. X-ray diffractogram of the (**a**) *Sample #1—Quartz sandstone.* Minerals: quartz, calcite, kaolinite (inset), (**b**) *Sample #2—Basalt (Spilite)* Minerals: albite (plagioclase feldspar), epidote, clinochlore, actinolite, phillipsite, (**c**) *Sample #3—Sedimentary rock.* Minerals: quartz, calcite, kaolinite, muscovite, clinochlore, (**d**) *Sample #4—Bioclastic limestone.* Minerals: quartz, calcite, orthoclase.

Sample #3 (Figure 1c) is a sandstone rock containing grains of detrital quartz (SiO$_2$), embedded in a matrix of carbonates, clay minerals and iron oxides/hydroxides. Due to its reddish-brown color and its composition, the sample could be a sandstone (formed in a continental environment). X-ray diffraction analysis revealed the presence of mainly quartz and calcite (CaCO$_3$), and small amounts of kaolinite (Al$_2$(Si$_2$O$_5$)(OH)$_4$), muscovite (KAl$_2$(AlSi$_3$O$_{10}$)(OH)$_2$), clinochlore (Mg$_5$Al(AlSi$_3$O$_{10}$)(OH)$_8$) (Figure 2c).

Sample #4 (Figure 1d) identified as extraclastic bioclastic limestone and could be described as a bioclastic grainstone. The extraclasts are represented mainly by angular–slightly rounded quartz and orthoclase (KAlSi$_3$O$_8$—potassium feldspar, Figure 1d) grains, with a micritic-microsparitic cement (calcite crystals are 5–10 μm in size). The fossil remnants consist mainly of foraminifera fragments (benthic and planktonic).

XRD measurements were performed on powders obtained from various areas of the samples. The obtained X-ray diffractograms are presented in Figure 2a–d and the identified crystalline structures are listed in Table 1. These results confirm the presence of the previously mentioned minerals in all samples. This confirmation allows us to have a better understanding of the interactions between the laser beam and the target and how the overall ablation process is affected by the presence of such wide spread of minerals.

Table 1. Minerals confirmed by the X-ray diffraction (XRD) data base and their indicative.

Rocks	Confirmed Minerals from XRD Database
Sample #1	PDF 01-070-7244 SiO_2 Quartz PDF 00-058-2001 $(Al_2Si_2)_5(OH)_4$ Kaolinite 1A PDF 00-005-0586 $CaCO_3$ Calcite
Sample #2	PDF 00-073-2147 $Ca_2Fe_{0.33}Al_{2.67}Si_3O_{12}OH$ Ca Fe (Epidote) PDF 00-046-1427 $(K,Na)_2(Si,Al)_8O_{16}*4H_2O$ (Philipsite) PDF 00-019-0749 $Mg_5Al(Si_3Al)O_{10}(OH)_8$ (Clinochlore) PDF 00-001-0739 $NaAlSi_3O_8$ (albite) PDF 00-080-0521 $Ca_2(Mg,Fe)_5Si_8O_{22}(OH)$ (Actinolite)
Sample #3	PDF 01-070-7344 SiO_2 Quartz PDF 00-024-0027 $CaCO_3$ Calcite PDF 01-078-2110 $Al_4(OH)_8(Si_4O_{10})$ Kaolinite PDF 01-070-1869 $K_{0.77}Al_{1.93}(Al_{0.5}Si_{3.5})O_{10}(OH)_2$ Muscovite-2M2 PDF 00-007-0078 $(Mg,Fe,Al)_6(Si,Al)_4O_{10}(OH)_8$ Clinochlore
Sample #4	PDF 01-083-1762 $Ca(CO_3)$ Calcite PDF 01-070-7344 SiO_2 Quartz PDF 01-083-1324 $K_{0.59}Ba_{0.19}Na_{0.33}(Al_{0.18}Si_{2.82}O_8)$ Orthoclase

The data base indicatives of the minerals which fit with the XRD peaks are shown in Table 1.

The chemical composition and distribution of the main elements were analyzed by EDX (Table 2). The results revealed the presence of metallic atoms in all samples which correspond to the elements found in the more complex minerals observed by XRD. However, for Sample #1, traces of Ti and Mg were noticed which can be considered as impurities due to the environment and natural conditions. Similar discussion can be made for all samples as all of them present a more heterogeneous distribution of elements on the surface.

Table 2. Results from EDX measurements depicting each sample elemental configuration and the respective error bar for each element.

Sample #1 (Petra)		Sample #2 (Bowen Island)		Sample #3 (Pico Anayet)		Sample #4 (Pico Collarada)	
Element	At %	Element	At %	Element	At %	Element	At %
Si	19.14 ± 0.25	Si	21.97 ± 0.25	Ca	31.85 ± 0.2	Ca	24.62 ± 0.3
Ca	5.78 ± 0.1	Fe	6.50 ± 0.13	Fe	1.33 ± 0.1	Si	17.98 ± 0.2
C	1.74 ± 0.1	Al	9.31 ± 0.16	C	0.27 ± 0.05	Al	2.84 ± 0.1
Al	2.14 ± 0.12	Mg	5.03 ± 0.11	Si	1.00 ± 0.08	Mg	0.93 ± 0.08
Ti	0.23 ± 0.05	Ba	0.68 ± 0.1	Al	0.6 ± 0.05	K	4.39 ± 0.1
Fe	0.68 ± 0.05	Ca	2.04 ± 0.11	Mn	0.67 ± 0.06	Fe	0.81 ± 0.06
Mg	0.53 ± 0.05	K	1.63 ± 0.08	K	1.47 ± 0.08	C	0.34 ± 0.05
O	69.71 ± 1.1	Na	0.73 ± 0.07	O	62.77 ± 1.2	O	48.06 ± 1.0
		Cl	0.58 ± 0.05				
		P	1.05 ± 0.07				
		C	1.13 ± 0.05				
		O	49.27 ± 1.0				

Two of the most important elements in the earth's crust are silicon (Si) and calcium (Ca), which are part of the various minerals (silicates, quartz, respectively calcite), so we focused on the concentrations of these two elements. From Table 2 it can be observed that Sample #2 contains the lowest concentrations of calcium (~6%) while Sample #3 contains the lowest concentration of Silicon (1%) and the highest concentration of calcium (~32%). Thus it is safe to assume that Sample #2 may contain very low concentration of calcite and Sample #3 contains low concentrations of silicates but high concentration of calcite. Sample #2 contains high concentrations of silicon (~18%), iron (~6.5%) and aluminum (~9%) which points to the inclusion of significant quantities of alumino-ferro-silicates.

For all investigated samples we performed an estimation analysis related to the micro porosity of the surface. We determined micro-porosities between 4 and 12% across all samples' surfaces (Sample #1—12%, Sample #2—4.7% Sample #3—9.23%, Sample #4—4.9%). The porosity was determined over an average of 10 surfaces of 1 mm^2, with the mention that throughout the surface of the samples the values of micro porosity were similar. These differences in the porosity of the targets could be expected to influence the values of the expansion velocities of the laser produced plasmas.

3.2. Optical Investigations of Laser Produced Plasmas

3.2.1. ICCD Fast Camera Imaging

When a laser beam impinges onto a surface the beam energy is absorbed by the target. For the case of homogenous materials, the energy is transferred to the electrons which are the first species ejected from the target by means of Coulomb explosion, while the rest of the target goes through various phase changes from solid–liquid–vapor, thus completing the ablated cloud which represents the object of study for LIBS. As expected for the case of our samples the beam energy will be absorbed differently by the various minerals presented in the target. These phenomena can lead to a complex ablation process which is harder to be analyzed than in the case of samples of single elements or even minerals.

The general LIBS technique is used for the identification of elements from a sample. More information (in terms of plume center-of-mass velocity and wave front dynamics) can be obtained by acquiring ICCD images at various delays. This approach was implemented in the study of simple [29] or complex targets [30], in controlled conditions.

The fast camera imaging is suitable for transient phenomena investigation. In the case of laser produced plasmas (LPP), it is necessary to have an adequate triggering system as LPP generally have a lifetime of a few μs [31]. Each recorded image is generally described by a series of parameters: resolution (which is given by the CCD detector and the optical system), time-delay (the moment of time, with respect to the trigger signal, at which acquiring starts), and the gate width (or integration time). In this case the initial moment (t = 0) is considered to be the "laser beam—target interaction moment". In order to have a good temporal resolution the gate width is usually of a few ns and it can increase towards longer evolution time where the plume is more rarefied and the emission is weaker. After their recording the images are transferred to the computer where they can be further analyzed. In order to estimate the expansion velocities, bi-dimensional images ("snapshots") of the laser-produced plasmas were recorded at constant laser fluence (19 J/cm^2) at various moments in time with respect to the laser beam (Figure 3). During the expansion, the LPP increases its volume and the center-of-mass, estimated as the maximum emission intensity zone, shifts towards higher distances as the recording time is changed. This leads to the conclusion that the expansion velocity is constant during the whole lifetime of the plume.

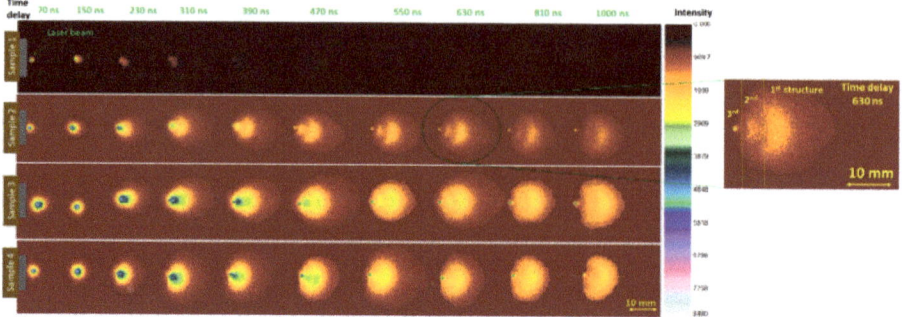

Figure 3. ICCD camera images, on a range of 1 μs, of laser produced plasmas generated by nanosecond laser ablation of mineral samples and a detail of the three-plasma structure (inset).

The cross-section on the expansion direction of the recorded images (Figure 4) shows more clearly the presence of two maxima which were attributed to two plasma components [32]. Due to the difference in their expansion velocities, in literature they can be found as the fast structure (or the "first structure") and slow structure (or the "second structure") [30]. Each of the two plasma structures expands with constant, yet different velocities. The velocity of each structure was determined by distance over time representation of the maximum intensity characteristic to each structure, method in line with the theoretical view of the laser produced plasma expansion at low pressures. The constant nature of the expansion is given by the linearity, respect for all the investigated plasmas.

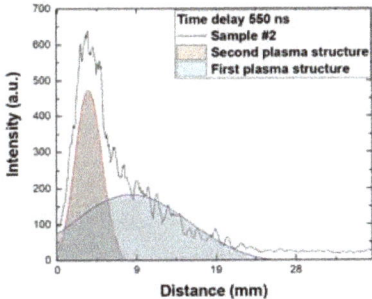

Figure 4. Cross section on the ICCD snapshot collected at 550 ns of the laser produced plasmas (LPP) generated on Sample #2.

The plume splitting behavior has been experimentally reported by several groups [33,34] and it is considered as a result of the different ejection mechanisms involved in the ns-laser ablation process. Therefore, the first—fast structure of the plume is ascribed to the electrostatic ejection mechanism (Coulomb explosion), while the second—slow structure corresponds to the thermal mechanisms (phase explosion, explosive boiling, evaporation). Our results show the presence of a third plasma structure (inset Figure 3), for samples #2, #3, and #4, described by a small emission region in the proximity of the target. In literature this structure is attributed to the presence of clusters, nanoparticles or molecules and it has its origin in the Knudsen layer which is usually characterized by black body radiation [35–37]. The structures observed in this study can be compared to the structures observed in ablation on Ni, Al, stainless-steel [28,29] or more complex targets like GeSe chalcogenide glasses [30].

For the samples #1 and #2 the expansion velocities of the first structure of the order of tens of km/s (Sample #1—11 km/s, Sample #2—13.5 km/s) while for the second structure we found velocities the order of a few km/s (Sample #1—4 km/s and Sample #2—6.5 km/s).

Although the velocity of second structure of the plasma plume generated on Sample #3 and #4 is in line with the values of the previous samples (Sample #3—3 km/s and Sample #4—6 km/s), for the first structure we found relative low velocities (Sample #3—8 km/s and Sample #4—9 km/s) most probable related to the nature of the mineral in each sample which could enhance the thermal ablation mechanism in the detriment of the electrostatic ones. This aspect of the laser-produced plasma is consistent with other reported results on pure metals or other complex materials [30] and it is important for the LIBS techniques as the most part of the emission is given by this second thermalized structure. For the second structure, in the case of nanosecond laser ablation, the emission is enhanced by the absorption of the laser beam tail by the ejected particle cloud.

With respect to the expansion velocity values of the first and second plasma structures, the highest ones were found for the plasma generated on Sample #2 (Figure 5), the one that contains almost no Calcite. The lowest velocities were observed for the plasma plumes of Sample #3—the one that contains very low concentrations of Quartz. All the other plasmas (generated on sample with both Calcite and Quartz) were found to be expanding with intermediate velocities. We note however that no correlation was observed between the porosity of the target and the estimated velocities. Samples

#2 and #4 presented similar porosities but strong differences in expansion velocities for each of the two plasma structures. A similar observation can be made for Samples #3 and #1 where for approximately similar porosities, consistent differences in the expansion velocities were observed.

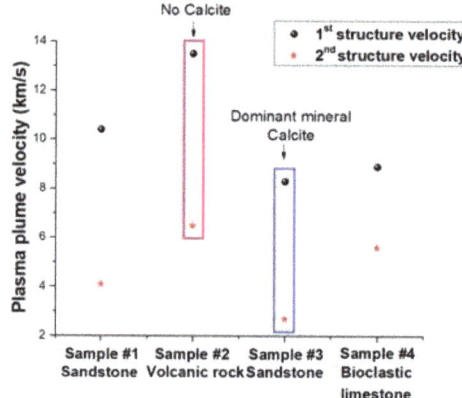

Figure 5. Plasma structure velocities for Sample #1, #2, #3, and #4 and their dependence on the abundance of calcite structure.

A possible explanation for this behavior can result by analyzing the energy transfer during laser-target interaction. For samples where, strong bonds like C=O or Ca-O are present, more of the incident laser energy is used on breaking the bond, leading to some relatively slower plasmas. This is the case for Sample #1, #3, and #4. These differences can be seen as a signature of the petrographic origins of the investigated rocks: Sample #2 is part of the magmatic rock family, while the other three samples belong to the sedimentary family—Sample #1 and Sample #3 are sandstones while Sample #4 is a bioclastic limestone.

3.2.2. Optical Emission Spectroscopy

The Optical Emission Spectroscopy technique [38] can help to determine the nature of the ejected particles through the energetic levels by identifying the wavelength and by using specialized databases [39]. The profile and intensity of the spectral lines can also provide information regarding the interactions between the ejected particles (e.g., Stark broadening [40]) and the internal energy of the plasma (i.e., electron temperature and electron density). For the plasma generated on each target, the global emission spectra were collected using a gate width of 2 µs. The experimental configuration ensures the collection of 600 µm plasma slice centered on the main expansion direction, thus providing a global characteristic in both a spatial and temporal perspective. This was done in order to collect all the emission lines regardless of their flight time [31]. For each of the sample we have identified atomic and ionic species characteristic for Ca, Si, Al, Mg, C, and O. Samples #2 and #3 have revealed the presence of their elements like K, Cl or S (see Figure 6). Most of the emission lines correspond to the elements identified with the EDX and XRD techniques, as discussed in the previous section. Thus, a qualitative comparison can be done between the LIBS signal and the EDX measurements. We notice that the evolution trends observed from EDX follow the LIBS signal changes, as such the increase in Ca amount by a factor of 17 in the target would lead to an enhancement of the Ca line intensity by a factor of 19. For the case of Si, an increase by a factor of 26 would only lead to an increase of a factor of 6. Finally, for all the other elements which were fund in significant smaller amounts (such as Al or Fe) an 8 times higher concentration would lead to an increased emission lines intensity of approximately 8 times. We observe that although the ratio is not always kept, especially for Si or lighter elements like K, most of the elements follow the changes from the target. At this moment, a quantitative proportionality is difficult the be achieved between the LIBs signal and the EDX data, given the complex process involved in both investigation techniques. However, under LTE the LIBS line intensity depends on the concentration of neutrals and singly ionized species, which allow us to tentatively use a qualitative comparison between the two techniques.

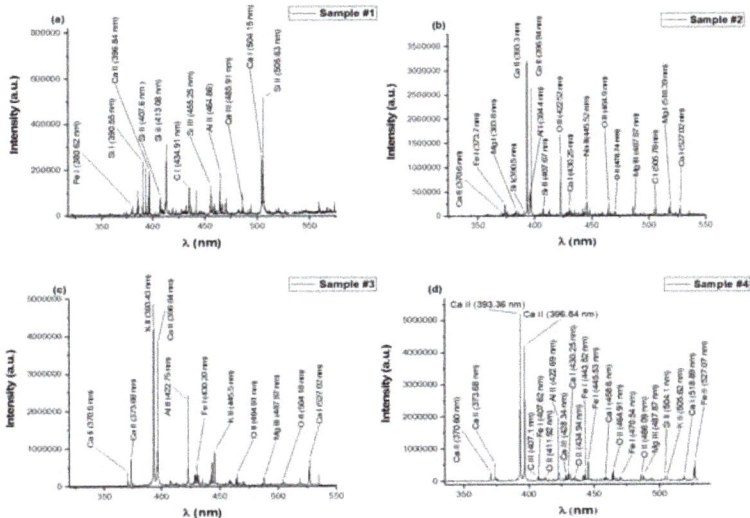

Figure 6. Global emission spectra collected at 1 mm from the target with a 2 μs gate width, gate delay of 100 ns, and laser fluence of 19 J/cm^2 of all the investigated samples (Sample #1 (**a**), Sample #2 (**b**) Sample #3 (**c**), Sample #4 (**d**)).

Most if not all plasma diagnostics techniques are valid under the assumption of the presence of a local thermodynamic equilibrium (LTE). However, in the case of transient plasmas all the plasma parameters such as electron temperature and particle density have a steep decrease in both time and space and thus the equilibrium has to be understood in a dynamic mode. There are different approached to estimate the limit of LTE with the most common one being the McWhirter criterion [40]:

$$N_e(\text{cm}^{-1}) \geq 1.6 \times 10^{12} \Delta E^3 (eV) T_e^{\frac{1}{2}}(K)$$

The above relation provides a real threshold above which we can assume LTE and implement the investigation techniques to further determine the excitation temperatures. Particularly we found 1.23×10^{15} cm^{-3} for Sample #1, 1.5×10^{15} cm^{-3} for Sample #2, 1.25×10^{15} cm^{-3} for Sample #3 and finally 1.18×10^{15} cm^{-3} for Sample # 4.

Once established the limit for which LTE model can be applied, the electron density can be estimated from the Saha-Eggert equation [41]. The relationship connects the plasma ionization equilibrium temperature to the proportion of population of two successive ionization states:

$$n_e = 4.83 \cdot 10^{15} \frac{I^* g^+ A^+ \lambda^*}{I^+ g^* A^* \lambda^+} T_e^{1.5} e^{-\frac{V^+ + E^+ - E^*}{k_B T_e}}$$

where the (*, +) superscripts represent the neutral excited atom and the singly charged ion, respectively, I is the emission intensities of a spectral line of λ wavelength (nm), T is the ionization temperature (expressed in K), which is taken as the excitation temperature in LTE conditions, V^+ is the first ionization potential, and E is the energy of the upper level of the transition. By implementing the Saha-Eggert equation we found a series of global n_e values for our plasmas that characterizes the 600 μm wide plasma volume throughout its evolution: 1.5×10^{16} cm^{-3} for Sample #1, 5.5×10^{16} cm^{-3} for Sample #2,

8.2×10^{16} cm^{-3} for Sample #3 and respectively 1.6×10^{17} cm^{-3} for Sample #4. The electron density exceeds the LTE limit with about one order of magnitude, thus, within our experimental conditions of laser fluence, background pressure and acquisition parameters, all the investigated plasmas respect the McWhirter criterion for LTE.

The excitation temperature of the plasma can be simply calculated from [41] using the intensity ratio of two spectral lines characterizing the same species (ion or atom), with the specific spectroscopic data (E, A, f) can be found in various databases (e.g., [39]). We note however that there are some reserves regarding the latter parameters, which can lead to significant uncertainties. In order to minimize the errors regarding the values of the oscillator strengths, it is suitable to use not two but a series of atomic lines with different upper excitation levels. The Boltzmann plot method represents the logarithmic function of the line intensity versus the upper level energy:

$$\ln\left(\frac{I_{ki}\lambda}{g_k A_{ki}}\right) = \ln\left(N_0 \frac{hc}{4\pi Z(T)}\right) - \frac{E_k}{k_b T_e}$$

The slope of this representation will give the excitation temperature, and its linearity or the deviation from it can be considered as an indication of LTE validity (an example can be seen in Figure 7).

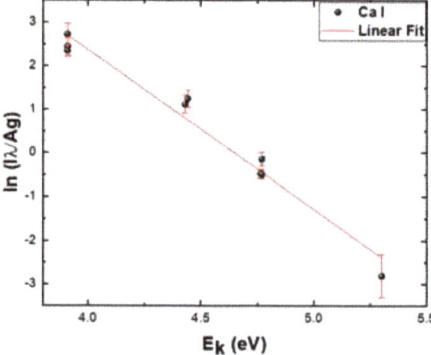

Figure 7. Representative Boltzmann plot for the Ca atoms representing the plasma generated on Sample #2.

The values of the excitation temperature were found to be in a range of 0-1 eV for the atomic species (i.e., Ca in Sample #1-6264 K, Sample #2-9976 K, Sample #3-6496 K and Sample #4-5800 K) and of about one order of magnitude higher for the ions (i.e., Ca in Sample #1-39440 K, Sample #2-15080 K, Sample #3-32480 K and Sample #4-41760 K). These discrepancies were previously reported by our group in [30,42,43] were they were related to the differential heating of the plume by the incoming laser beam. However, in the LTE conditions we would expect that regardless of the nature of the atom/ion investigated the plasma temperature should have the same values. For all the investigated atoms the values of the temperatures are almost the same (within a 5% error margin), while for the ions the discrepancies are higher. For the identified metallic species (Fe, Al, Ti, Mg) in the plasmas we found electron temperature between 3480 K and 5800 K through all the investigated plasmas while for lighter elements like Ca, C, Si or K we found significantly increased temperature (from 6264 K for Ca up to 19,720 K for K). The excitation temperatures were found to be relatively close to the ones reported for laser produced plasmas on copper (13,200 K to 17,200 K) or lead (11,700 K to 15,300 K) [44], while for studies on a Silicon plasma [45] it was reported that the electron temperatures varied between 6000 K to 9000 K.

The plasmas generated on Sample #1, #3, and #4 (all sedimentary rocks) present the highest values for the excitation temperature for the ionic species of Ca and Si, with respect to the other investigated plasma. This is in line with the structure observed through EDX investigation, as all three targets have a significant larger concentration of Ca. The result also correlates well with the values of the expansion velocity of the first plasma structure, determined through ICCD fast camera imaging. On the other hand, the plume generated on Sample #2 (volcanic rock) presents the highest values for the neutral species, having the highest expansion velocity for the second plasma structure. In literature [30,42,43] the first plasma structure contains mainly ions, while the second and the third one contains mainly neutral species. This view over the nature of the plasma components it is also confirmed by our experimental results. The spectral investigations revealed that the temperature for the ionic Ca is much lower than the ones of ionic Ca for the other samples. This could be related to the nature of Sample #2—a magmatic rock that has almost no Calcite in it versus the other samples which are sandstones.

The experimental data showcased that the properties of the LPP strongly depend on the nature of the rocks. Nevertheless, the nature of the rocks includes multiple variables as composition of minerals, elemental composition of each mineral, physical properties of the rock resulted from the mechanisms of mineral formation across time which were addressed in our study by performing XRD and Optical microcopy. However, not only the quantity of Calcite microcrystals are directly responsible for low velocity plasma plumes as there are other aspects to be considered like porosity of the rock, degree of impurity on the sample, crystallinity, grain size, hardness, coherence, and moisture or reflectivity of the target. The state of the sample is one major aspect that needs to be considered in order to have a better control on the LIBS process. Thus, the extension towards quantitative analysis in terms of atomic or ionic temperature and expansion velocity becomes a strong tool in material investigations and understanding how the history of the target affects the laser matter interaction process and subsequently the LIBS technique.

4. Mathematical Model

4.1. Fractal Analysis

Ablation plasma behaves like a fractal medium taking into account the collision processes amongst the ejected particles. Indeed, between two successive collisions, the trajectories of the plasma particles are straight lines, that become nondifferentiable in their impact points. Considering that all the collision impact points are an uncountable set of point, it results that the trajectories of the plasma particles become continuous but non-differentiable curves (fractal curves). Since in these conditions the non-differentiability (fractality) appears as a fundamental property of the ablation plasma dynamics it seems necessary to construct a corresponding non-differentiable plasma physics model, for example in the form of fractal hydrodynamics model [46]. The mathematics behind this model as well as some applications of the model were development in a systematic manner by our group in [28,29,46,47]. In the following, a fractal analysis will prove a multi-structuring of the ablation plasma in the form of Coulomb, thermal and cluster structures. Let us consider the solutions for the fractal hydrodynamic equations system in the following form [48]:

$$V = V_D + iV_F = \frac{v_0 \alpha^2 + \left(\frac{\lambda}{\alpha}\right)^2 xt}{\alpha^2 + \left(\frac{\lambda}{\alpha}\right)^2 t^2} + i\lambda \frac{(x - v_0 t)}{\alpha^2 + \left(\frac{\lambda}{\alpha}\right)^2 t^2} \quad (1)$$

$$\rho = \frac{1}{(\pi)^2 \left[\alpha^2 + \left(\frac{\lambda}{\alpha}\right)^2 t^2\right]^{1/2}} \exp\left[-\frac{(x - vt)}{\alpha^2 + \left(\frac{\lambda}{\alpha}\right)^2 t^2}\right] \quad (2)$$

$$\lambda = \mu (dt)^{\left(\frac{2}{D_F} - 1\right)}, i = \sqrt{-1} \quad (3)$$

In Equations (1)–(3) x is the fractal space coordinated, t is the non-fractal temporal coordinate, having the role of an affine parameter of the movement curve, V is the complex velocity field, V_D is the real component, independent on the scale resolution dt,

$$V_F = -\lambda \ln \rho$$

is the imaginary component dependent on the scale resolution, ρ is the state density, μ is a coefficient associated to the fractal-non-fractal transition, V_0 is the initial velocity, α is the Gaussian parameter characterizing the initial energy distribution and D_F is the fractal dimension of the movement curves. For the fractal dimension we can choose and use either Kolmogorov relation, either Hausdorff–Besikovici relation etc. [49,50]. Once chosen it needs to be constant and arbitrary: $D_F < 2$ for corelative processes, $D_F > 2$ for non-corelative processes, etc. [49]. We would like to remind that through the variable x, t, V, ρ and through parameters: α, μ, dt, D_F and V_0 we will be able to showcase various dynamics of multiple ablation plasma structures.

The solutions (1) and (2) can be simplified through the normalizations:

$$\frac{x}{\alpha} = \xi, \ \frac{V_{0}t}{\alpha} = \tau, \ \frac{V_D}{V_{D0}} = \overline{V}_D, \ \frac{V_F}{V_{F0}} = \overline{V}_F, \ \frac{\rho}{\rho_0} = \overline{\rho}, \ \left(\frac{\lambda}{\alpha V_0}\right) = \theta, \ V_0 = V_{D0}, \ \frac{\lambda}{\alpha} = V_{F0}, \ \rho_0 = \frac{1}{\alpha\sqrt{\pi}} \quad (4)$$

Then $\overline{V}_D$, $\overline{V}_F$, $\overline{\rho}$ become:

$$\overline{V}_D = \frac{1 + \theta^2 \xi \tau}{1 + \theta^2 \tau^2} \quad (5)$$

$$\overline{V}_F = \frac{\theta(\xi - \tau)}{1 + \theta^2 \tau^2} \quad (6)$$

$$\overline{\rho} = \frac{1}{(1 + \theta^2 \tau^2)^{1/2}} \exp\left[-\frac{(\xi - \tau)^2}{1 + \theta^2 \tau^2}\right] \quad (7)$$

From Equations (5) and (7) the state current density at differentiable scale resolution takes the form:

$$\overline{J}_D = \overline{\rho}\overline{V}_D = \frac{1 + \theta^2 \xi \tau}{(1 + \theta^2 \tau^2)^{3/2}} \exp\left[-\frac{(\xi - \tau)^2}{1 + \theta^2 \tau^2}\right] \quad (8)$$

while, the state current density at fractal scale resolution takes the form:

$$\overline{J}_F = \overline{\rho}\overline{V}_F = \frac{\theta(\xi - \tau)}{(1 + \theta^2 \tau^2)^{3/2}} \exp\left[-\frac{(\xi - \tau)^2}{1 + \theta^2 \tau^2}\right] \quad (9)$$

During the expansion of a laser produced plasma we can identify three important moments (chronologically): The Coulomb explosion moment, the thermal ejection moment and the cluster formation moment. Each of these moments are defined by three different types of ejection mechanism which lead to the formation of three independent plasma structures [34,36,42,43]. In such a context the dynamics of fast plasma structure generated though Coulomb explosion mechanism would be described by relations (5), (7) and (8) while the dynamics of the slow structure generated through thermal ejection mechanisms are given by relations (6), (7) and (9). The reasoning behind this association is given by the fact that the nondifferentiable behavior of the laser ablation plasmas is induced through the collision process between the ejected particles in each plasma structure. In Figure 8 we have shown the 3D representation of states current density at differentiable and fractal resolution scale for different degrees of fractalizations, which is identified with the real particle density. With the increase of the fractalization degree we observe a change in the slope defining the velocity of the differentiable part of the current. This can be read as an increase in the thermal velocity of the particle as this component is induced by thermal mechanism. On the other hand, the particle current at a fractal resolution scale, induced by Coulomb mechanisms, will present two components. These components are generated by a double layer formed in the initial stages of ablation. The fractality degree has little contribution to the spatio–temporal evolution of this component, although at a higher resolution scale we observe a better separation between the two components of the current.

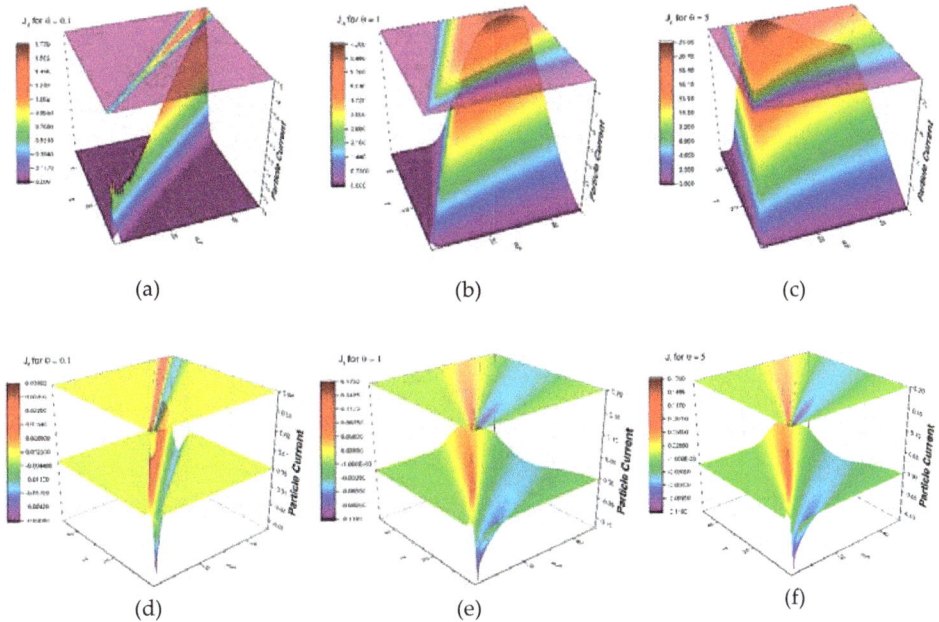

Figure 8. 3D representation and contour plot of the differentiable (**a**–**c**) and fractal (**d**–**f**) particle density for various degrees of fractalization ($\theta = 0.1, 1, 5$).

In order to describe the dynamics of the third substructure, containing mainly clusters and nanoparticles, we will postulate that the specific momentum at the global scale resolution is null. This means that at the differentiable scale resolution the velocity is equal and of opposite sign with the velocity at a fractal scale resolution.

$$V_D = -V_F = \lambda (dt)^{(\frac{2}{D_F}-1)} \partial_x (\ln \rho) \tag{10}$$

In these conditions the conservation law of the state density:

$$\partial_t \rho + \partial_{xx}(\rho V_D) = 0$$

takes the form of a fractal diffusion equation:

$$\partial_t \rho = \lambda (dt)^{(\frac{2}{D_F}-1)} \partial_{xx} \rho \tag{11}$$

The solution of this equations has the following expression [40]:

$$\rho(x,t) = \frac{a}{\left(4\pi\lambda(dt)^{(\frac{2}{D_F}-1)}t\right)^{1/2}} \exp\left[-\frac{(x-b)^2}{4\lambda(dt)^{(\frac{2}{D_F}-1)}t}\right] \tag{12}$$

where a and b are integrationn constant. In such a context the velocity takes the form:

$$v = \frac{x-b}{2t} \qquad (13)$$

while the states current density is:

$$j = \frac{a(x-b)}{\left(16\pi\lambda(dt)^{\left(\frac{2}{D_F}-1\right)}\right)^{1/2} t^{3/2}} \exp\left[-\frac{(x-b)^2}{4\lambda(dt)^{\left(\frac{2}{D_F}-1\right)} t}\right] \qquad (14)$$

Now, we calibrate the structure that contains mainly clusters with the dynamics of the other two structures in order to admin a normalization by imposing the restrictions: $a \equiv 1$ and $b \equiv 0$. We can find:

$$\bar{\rho} = \frac{1}{(4\theta\tau)^{1/2}} \exp\left[-\frac{\xi^2}{4\theta\tau}\right] \qquad (15)$$

$$\bar{V} = \frac{V_D}{V_0} = \frac{\xi}{2\tau} \qquad (16)$$

$$\bar{J} = \frac{\xi}{(4\theta\tau)^{1/2} \tau^{3/2}} \exp\left[-\frac{\xi^2}{4\theta\tau}\right] \qquad (17)$$

In Figure 9 we have represented the 3D representation of current density for various degrees of fractalization depicted through θ. The three ranges of values were chosen for the fractalization degree in order to cover the full range of ablation mechanism seen experimentally: Coulomb explosion, thermal evaporation, and explosive boiling. The range of fractality degrees which are specific for each ablation mechanism were defined in our previous work. Here we confirm that the range remains constant regardless of the nature of the target.

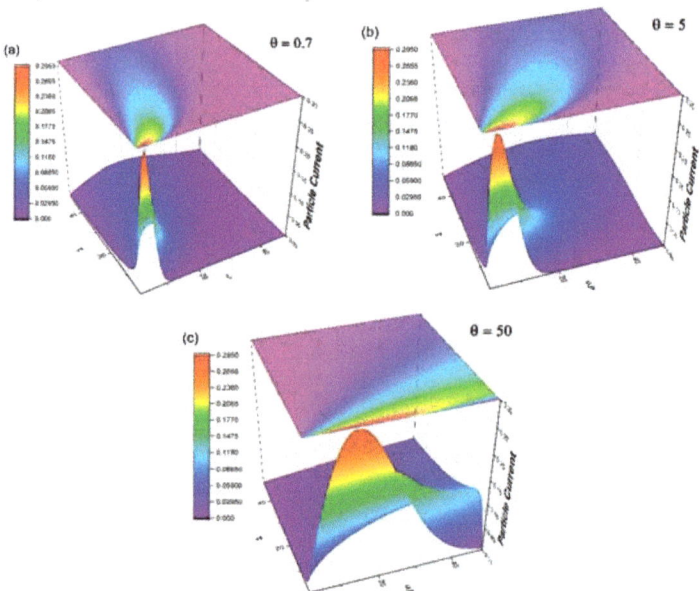

Figure 9. 3D representation and contour plot of the global particle density for various degrees of fractalization (θ = 0.7 (**a**), 5 (**b**) and 50 (**c**)).

In Figure 9 we see the space-time representation of the particle current density. The contour plot representation attached to the 3D representation shows that the maxima shifts during expansion. This behavior was seen experimentally through ICCD fast camera imagining (Figure 3). As the current maxima characteristic for each ablation mechanism shifts, it defines a unique slope which describes the expansion velocity of each structure. For the particles ejected through Coulomb explosions which are described by a low degree of fractalization a steep slope is defined and thus a high expansion velocity. The particles are mainly visible in the first moments of expansion. For those ejected through thermal mechanism we notice a different slope with a longer life-time and a bigger spatial expansion (characteristics of a reduced expansion velocity). The last structure is formed mainly by nanoparticles and clusters and it is defined by a high fractalization degree. We notice that the maximum of the particle current holds its value at small distances even for a long expansion time. This is a well-known trait of this complex structure. If we perform a small calculation by using the initial conditions of our experiments and impose them onto the fractal analysis we can estimate the expansion velocities of each plasma structure. We find for the first structure velocities of 18.7 km/s, for the second structure 2.5 km/s, while for the last structure 710 m/s. The results are in line with the values reported in literature. Thus, the fractal analysis is a robust one and can be implemented on a wide range plasma regardless the nature of targets.

4.2. Group Invariance Analysis

In the one-dimensional stationary case, the fractal hydrodynamic equation system becomes [48]:

$$\partial_{xx} u + k_0^2 u = 0 \tag{18}$$

where

$$u = \rho^{1/2}, \quad k_0^2 = \frac{E}{2m_0 \lambda^2} \tag{19}$$

E is an integration constant having a signification of the plasma entity energy and m_0 the rest mass of the entity.

The general solution of Equation (18) takes the following form [50]:

$$u = h e^{i(k_0 x + \Phi)} + \bar{h} e^{-i(k_0 x + \Phi)} \tag{20}$$

where h is a complex amplitude, $\bar{h}$ is the complex conjugate of h and Φ is the initial phase. Thus, h and $\bar{h}$ and Φ label each entity of the ablation plasma structure that has a general characteristic the Equation (18) and consequently the same k_0.

In such a context, can an a priori connection between the entities of the ablation plasma be established? Since (18) has a "hidden" symmetry in the form of a homographic group, we can answer to this question positively. Indeed, the ratio of two independent linear solution of (18), τ_0, is a solution of Schwartz differential equation [51]:

$$\{\tau_0, x\} = \frac{d}{dx}\left(\frac{\ddot{\tau}_0}{\dot{\tau}_0}\right) - \frac{1}{2}\left(\frac{\ddot{\tau}_0}{\dot{\tau}_0}\right)^2 = 2k_0^2 \tag{21}$$

$$\dot{\tau}_0 = \frac{d\tau_0}{dx}, \quad \ddot{\tau}_0 = \frac{d^2\tau_0}{dx^2}$$

The left part of Equation (21) is invariant with respect to homogeneous transformation:

$$\tau_0 \leftrightarrow \tau_0' = \frac{a_1 \tau_0 + b_1}{c_1 \tau_0 + d_1} \tag{22}$$

with a_1, b_1, c_1, d_1 real parameters. The set of transformations (23) corresponding to all possible values of the parameters is the group SL (2R). Therefore, the ablation plasma structure of all the entities having the same k_0 is in the bi-univocal correspondence with the transformation of the group SL (2R). This allows the construction of a "personal" parameter τ_0 for each entity of the ablation plasma structure separately. Indeed, we choose as "guide" the general form of the solution of (21) which is written in the form:

$$\tau'_0 = m + n \tan(k_0 x + \Phi) \tag{23}$$

where m, n and Φ are constant and characterizes an entity of the ablation plasma structure. By identifying the phase from Equation (23) with one from Equation (20) we can write the "personal" parameter of the entity as:

$$\tau'_0 = \frac{\overline{h}\tau_0 + h}{\tau_0 + 1}, \quad h = m + in, \overline{h} = m - in, \quad \tau_0 \equiv e^{-2i(k_0 x + \Phi)} \tag{24}$$

The fact that (24) is also a solution of Equation (21) implies, by explicating Equation (22) the group of transformation [51]

$$h' = \frac{a_1 h + b_1}{c_1 h + d_1}$$

$$k' = \frac{c_1 \overline{h} + d_1}{c_1 h + d_1} k \tag{25}$$

This group work as a group of "synchronization" among the various entities of the ablation plasma structure, process to which the amplitudes and phases of each of them obviously participate, in the sense that they are also connected. More precisely, by means of the group (25), the phase of k is only moved with a quantity depending on the amplitude of the entity of the ablation plasma structure at the transition among various entities of the ablation plasma structures. But not only that, the amplitude of the entity of the ablation plasma structure is also affected from a homographic perspective. The usual "synchronization" manifested through the delay of the amplitudes and phases of the entities of the ablation plasma structure must represent here only a totally particular case.

The infinitesimal generator of the group (25)

$$\hat{L}_1 = \frac{\partial}{\partial h} + \frac{\partial}{\partial \overline{h}}, \quad \hat{L}_2 = h\frac{\partial}{\partial h} - \overline{h}\frac{\partial}{\partial \overline{h}}, \quad \hat{L}_3 = h^2 \frac{\partial}{\partial h} + \overline{h}^2 \frac{\partial}{\partial \overline{h}} + (h - \overline{h})k\frac{\partial}{\partial k} \tag{26}$$

satisfies the commutation relations:

$$\left[\hat{L}_1, \hat{L}_2\right] = \hat{L}_1, \left[\hat{L}_2, \hat{L}_3\right] = \hat{L}_3, \left[\hat{L}_3, \hat{L}_1\right] = -2\hat{L}_2 \tag{27}$$

Thus, the structure of the group (25) is given by Equation (27) so that the only non-zero structure constants should be:

$$C^1_{12} = C^2_{23} = -1, C^2_{31} = -2 \tag{28}$$

Therefore, the invariant quadratic form is given by the "quadratic" tensor of the group (25)

$$C_{\alpha\beta} = C^\mu_{\alpha\nu} C^\nu_{\beta\mu} \tag{29}$$

where summation over the repeated indices is understood. Using Equations (28) and (29), the tensor $C_{\alpha\beta}$ is:

$$C_{\alpha\beta} = \begin{pmatrix} 0 & 0 & -4 \\ 0 & 2 & 0 \\ -4 & 0 & 0 \end{pmatrix} \tag{30}$$

meaning that the invariant metric of the group (25) has the form:

$$\frac{ds^2}{f} = \Omega_0^2 - 4\Omega_1 \Omega_2 \tag{31}$$

with f an arbitrary constant factor and

$$\Omega_0 = -i\left(\frac{dk}{k} - \frac{dh + d\bar{h}}{h - \bar{h}}\right), \Omega_1 = \frac{dh}{(h - \bar{h})k}, \Omega_2 = -\frac{kd\bar{h}}{h - \bar{h}} \quad (32)$$

three differential 1-forms, absolutely invariant through the group (25). In these conditions, the metric (31) becomes:

$$\frac{ds^2}{f} = -\left(\frac{dk}{k} - \frac{dh + d\bar{h}}{h - \bar{h}}\right)^2 + 4\frac{dhd\bar{h}}{(h - \bar{h})^2} \quad (33)$$

A particular case of ablation plasma entities synchronization is the one induced through the parallel transport of direction in a Levi-Civita sense [51]. Then, in the space of variables $(h, \bar{h}, k)$ the differential 1-form Ω_0 is null, $\Omega_0 = 0$, or in the space of variables (m, n, Φ):

$$d\Phi = -\frac{dm}{n} \quad (34)$$

In such a situation the metric (33) can be reduced to the Lobacewski plane metric in Poincare representation:

$$\frac{ds^2}{f} = \frac{dhd\bar{h}}{(h - \bar{h})^2} = \frac{dm^2 + dn^2}{n^2} \quad (35)$$

Then the functionality of a variational principle of Matzner-Misner type associated to the Lagrangian built with the metric (35) implies the "field equations" [51,52]:

$$(h - \bar{h})\nabla^2 h = 2\nabla h \nabla h$$

$$(h - \bar{h})\nabla^2 \bar{h} = 2\nabla \bar{h} \nabla \bar{h} \quad (36)$$

The solution are as follows:

$$h = m + in = -i\frac{\cos h\Psi - e^{-i\chi}\sin h\Psi}{\cos h\Psi + e^{-i\chi}\sin h\Psi} \quad (37)$$

with

$$\nabla^2 \Psi = 0 \quad (38)$$

and χ real and arbitrary, which means that for m and n we have the following expression:

$$m = \frac{2r\sin\Phi}{1 + r^2 + 2r\cos\Phi}, n = \frac{1 - r^2}{1 + r^2 + 2r\cos\Phi} \quad (39)$$

$$r = \tanh\Psi$$

Admitting that for χ a continuous variation, the solution (20) with the restriction (34) implies the synchronization of the ablation plasma structure entities in the form of self-modulation in amplitude through transformations of Stoler type [48,51]. Taking into consideration the meaning of such a transformation it results that the self-modulation in amplitude is realized through "annihilation" and "creation" of charges, i.e., through recombinations and ionizations.

5. Conclusions

Various samples with different geomorphological backgrounds collected from the Northern Hemisphere were investigated. Optical microscopy and XRD analysis were used to identify the minerals present in the rocks while EDX measurements revealed the composition of the samples.

The dynamics of laser induced plasmas on the four samples were investigated by means of ICCD fast camera imaging and optical emission spectroscopy. The images showed for three samples the split of the ejected cloud into three distinct structures. Optical emission spectroscopy allowed the identification of all the elements from the target and confirmed the results obtained by EDX and XRD. The Saha-Eggert equation was used to determine the electron density and the McWhirter criterion was used to verify if the local thermodynamic equilibrium conditions were met. All the investigated plasmas presented electron densities above the LTE threshold. The Boltzmann plot method was used to determine the excitation temperatures. Important differences were found between the values of each species and their corresponding ions, result discussed in the framework of selective heating by the incoming pulse.

The values of the global plasma structure and those of the individual species temperatures, were connected to the presence of calcite in the samples. Other aspects like target porosity were found to play a smaller role in the plasma plume dynamic. The samples with highest velocities were presenting low concentration of calcite while the sedimentary rocks were presenting low expansion velocities and high excitation temperatures, in line with effects of a strong thermalized ablation process.

A mathematical model to describe the dynamics in ablation plasma using various mathematical operational procedures is developed: multi structuring of the ablation plasma, by means of the fractal analysis and synchronizations of the ablation plasma entities through SL (2R) group invariance and in a particular case through self-modulation in the form of Stoler type transformations. The model proves that during the synchronization phenomena not only the amplitudes but also the phases are affected from a homographic perspective. A self-modulation through a Stoler type transformation is initiated functionalizing charge creating/annihilation processes.

Author Contributions: Conceptualization, S.G., M.A. and G.B.; methodology. G.B., H.B., S.G., N.C. and M.A.; investigation, S.A.I., F.E., N.C., H.B., M.A. and S.G.; writing—original draft preparation, S.A.I., F.E., H.B., G.B.; writing—review and editing, S.A.I., F.E. and M.A.; visualization, S.A.I.; supervision, M.A. and S.G.

Funding: This work was financially supported by Romanian Space Agency (ROSA) within the STAR Program, Project no.: 169/20.07.2017, by the National Authority for Scientific Research and Innovation in the framework of Nucleus Program–16N/2019 and by a grant of Ministry of Research and Innovation, CNCS-UEFISCDI, project number PN-III-P4-ID-PCE-2016-0355, within PNCDI III.

Conflicts of Interest: The authors declare no conflict of interest.

References

1. Davari, S.A.; Hu, S.; Mukherjee, D. Calibration-free quantitative analysis of elemental ratios in intermetallic nanoalloys and nanocomposites using Laser Induced Breakdown Spectroscopy (LIBS). *Talanta* **2017**, *164*, 330–340. [CrossRef] [PubMed]
2. Hahn, D.W.; Omenetto, N. Laser-Induced Breakdown Spectroscopy (LIBS), Part I: Review of Basic Diagnostics and Plasma–Particle Interactions: Still-Challenging Issues Within the Analytical Plasma Community. *Appl. Spectrosc.* **2010**, *64*, 335–366. [CrossRef] [PubMed]
3. Xu, T.; Liu, J.; Shi, Q.; He, Y.; Niu, G.; Duan, Y. Multi-elemental surface mapping and analysis of carbonaceous shale by laser-induced breakdown spectroscopy. *Spectrochim. Acta Part B-At. Spectrosc.* **2016**, *115*, 31–39. [CrossRef]
4. Xie, S.; Xu, T.; Han, X.; Lin, Q.; Duan, Y. Accuracy improvement of quantitative LIBS analysis using wavelet threshold de-noising. *J. Anal. At. Spectrom.* **2017**, *32*, 629–637. [CrossRef]
5. El Haddad, J.; Canioni, L.; Bousquet, B. Good practices in LIBS analysis: Review and advices. *Spectrochim. Acta-Part B At. Spectrosc.* **2014**, *101*, 171–182. [CrossRef]
6. Legnaioli, S.; Lorenzetti, G.; Pardini, L.; Cavalcanti, G.H.; Palleschi, V. Applications of LIBS to the Analysis of Metals. In *Laser-Induced Breakdown Spectroscopy*; Springer: Berlin/Heidelberg, Germany, 2014.
7. Eppler, A.S.; Cremers, D.A.; Hickmott, D.D.; Ferris, M.J.; Koskelo, A.C. Matrix effects in the detection of Pb and Ba in soils using laser-induced breakdown spectroscopy. *Appl. Spectrosc.* **1996**, *50*, 1175–1181. [CrossRef]

8. Sanghapi, H.K.; Jain, J.; Bol'Shakov, A.; Lopano, C.; McIntyre, D.; Russo, R. Determination of elemental composition of shale rocks by laser induced breakdown spectroscopy. *Spectrochim. Acta-Part B At. Spectrosc.* **2016**, *122*, 9–14. [CrossRef]
9. Popov, A.M.; Zaytsev, S.M.; Seliverstova, I.V.; Zakuskin, A.S.; Labutin, T.A. Matrix effects on laser-induced plasma parameters for soils and ores. *Spectrochim. Acta-Part B At. Spectrosc.* **2018**, *148*, 205–210. [CrossRef]
10. Barker, M.; Rayens, W. Partial least squares for discrimination. *J. Chemom.* **2003**, *17*, 166–173. [CrossRef]
11. Dyar, M.D.; Carmosino, M.L.; Breves, E.A.; Ozanne, M.V.; Clegg, S.M.; Wiens, R.C. Comparison of partial least squares and lasso regression techniques as applied to laser-induced breakdown spectroscopy of geological samples. *Spectrochim. Acta-Part B At. Spectrosc.* **2012**, *70*, 51–67. [CrossRef]
12. Harmon, R.S.; Russo, R.E.; Hark, R.R. Applications of laser-induced breakdown spectroscopy for geochemical and environmental analysis: A comprehensive review. *Spectrochim. Acta-Part B At. Spectrosc.* **2013**, *87*, 11–26. [CrossRef]
13. Tognoni, E.; Cristoforetti, G.; Legnaioli, S.; Palleschi, V.; Salvetti, A.; Müller, M.; Gornushkin, I. A numerical study of expected accuracy and precision in calibration-free laser-induced breakdown spectroscopy in the assumption of ideal analytical plasma. *Spectrochim. Acta-Part B At. Spectrosc.* **2007**, *62*, 1287–1302. [CrossRef]
14. Bulajic, D.; Corsi, M.; Cristoforetti, G.; Legnaioli, S.; Palleschi, V.; Salvetti, A.; Tognoni, E. A procedure for correcting self-absorption in calibration free-laser induced breakdown spectroscopy. *Spectrochim. Acta-Part B At. Spectrosc.* **2002**, *57*, 339–353. [CrossRef]
15. González, R.; Lucena, P.; Tobaria, L.M.; Laserna, J.J. Standoff LIBS detection of explosive residues behind a barrier. *J. Anal. At. Spectrom.* **2009**, *24*, 1123–1126. [CrossRef]
16. Waltham, T. The sandstone fantasy of Petra. *Geol. Today* **1994**, *10*, 105–111. [CrossRef]
17. Makhlouf, I.M.; Abed, A.M. Depositional facies and environments in the Umm Ishrin Sandstone Formation, Dead Sea area, Jordan. *Sediment. Geol.* **1991**, *71*, 177–187. [CrossRef]
18. Migoń, P.; Goudie, A. Sandstone geomorphology of south-west Jordan, Middle East. *Quaest. Geogr.* **2014**, *33*, 123–130. [CrossRef]
19. Amireh, B.S. Mineral composition of the Cambrian-Cretaceous Nubian Series of Jordan: Provenance, tectonic setting and climatological implications. *Sediment. Geol.* **1991**, *71*, 99–119. [CrossRef]
20. Leitch, H.C.B. Contributions to the Geology of Bowen Island. Ph.D. Thesis, University of British Columbia, Kelowna, BC, Canada, 1947.
21. Boyle, D.R.; Turner, R.J.W.; Hall, G.E.M. Anomalous arsenic concentrations in groundwaters of an island community, Bowen Island. British Columbia. *Environ. Geochem. Health* **1998**, *20*, 199–212. [CrossRef]
22. Friedman, R.M.; Monger, J.W.H.; Tipper, H.W. Age of the Bowen Island Group, southwestern Coast Mountains, British Columbia. *Can. J. Earth Sci.* **1990**, *27*, 1456–1461. [CrossRef]
23. Valero Garcés, B.L.; Aguilar, J.G. Shallow carbonate lacustrine facies models in the Permian of the aragon-bearn basin (Western Spanish-French Pyrenees). *Carbonates Evaporites* **1992**, *7*, 94–107. [CrossRef]
24. Sibuet, J.C.; Srivastava, S.P.; Spakman, W. Pyrenean orogeny and plate kinematics. *J. Geophys. Res.* **2004**, *109*, B08104. [CrossRef]
25. Dunham, R.J. Classification of Carbonate Rocks According to Depositional Textures. In *Classification of Carbonate Rocks—A Symposium*; Ham, W.E., Ed.; American Association of Petroleum Geologists: Vancuver, BC, Canada, 1962; pp. 108–121.
26. Roigé, M.; Gómez-Gras, D.; Remacha, E.; Daza, R.; Boya, S. Tectonic control on sediment sources in the Jaca basin (Middle and Upper Eocene of the South-Central Pyrenees). *C. R. Geosci.* **2016**, *348*, 236–245. [CrossRef]
27. Rodríguez, L.; Cuevas, J.; Tubía, J.M. Structural Evolution of the Sierras Interiores (Aragón and Tena Valleys, South Pyrenean Zone): Tectonic Implications. *J. Geol.* **2014**, *122*, 99–111. [CrossRef]
28. Irimiciuc, S.; Bulai, G.; Agop, M.; Gurlui, S. Influence of laser-produced plasma parameters on the deposition process: In situ space- and time-resolved optical emission spectroscopy and fractal modeling approach. *Appl. Phys. A Mater. Sci. Process.* **2018**, *124*, 615. [CrossRef]
29. Irimiciuc, S.A.; Mihaila, I.; Agop, M. Experimental and theoretical aspects of a laser produced plasma. *Phys. Plasmas* **2014**, *21*, 93509. [CrossRef]
30. Irimiciuc, S.; Boidin, R.; Bulai, G.; Gurlui, S.; Nemec, P.; Nazabal, V.; Focsa, C. Laser ablation of $(GeSe_2)_{100-x}$ $(Sb_2Se_3)_x$ chalcogenide glasses: Influence of the target composition on the plasma plume dynamics. *Appl. Surf. Sci.* **2017**, *418*, 594–600. [CrossRef]

31. Puretzky, A.A.; Geohegan, D.B.; Haufler, R.E.; Hettich, R.L.; Zheng, X.Y.; Compton, R.N. Laser ablation of graphite in different buffer gases. *AIP Conf. Proc.* **1993**, *288*, 365–374.
32. Amoruso, S.; Armenante, M.; Bruzzese, R.; Spinelli, N.; Velotta, R.; Wang, X. Emission of prompt electrons during excimer laser ablation of aluminum targets. *Appl. Phys. Lett.* **1999**, *75*, 7–9. [CrossRef]
33. Amoruso, S.; Unitá, I.; Fisiche, S.; Federico, N.; Angelo, M.S.; Cintia, V.; Napoli, I.; Toftmann, B.; Schou, J. Thermalization of a UV laser ablation plume in a background gas: From a directed to a diffusionlike flow. *Phys. Rev. E* **2004**, *69*, 56403. [CrossRef]
34. Harilal, S.S.; Bindhu, C.V.; Tillack, M.S.; Najmabadi, F.; Gaeris, A.C. Plume splitting and sharpening in laser-produced aluminium plasma. *J. Phys. D Appl. Phys.* **2002**, *35*, 2935–2938. [CrossRef]
35. Harilal, S.S.; Bindhu, C.V.; Tillack, M.S.; Najmabadi, F.; Gaeris, A.C. Internal structure and expansion dynamics of laser ablation plumes into ambient gases. *J. Appl. Phys.* **2003**, *93*, 2380–2388. [CrossRef]
36. Chen, Z.; Bogaerts, A. Laser ablation of Cu and plume expansion into 1 atm ambient gas. *J. Appl. Phys.* **2005**, *97*, 063305. [CrossRef]
37. O'Mahony, D.; Lunney, J.; Dumont, T.; Canulescu, S.; Lippert, T.; Wokaun, A. Laser-produced plasma ion characteristics in laser ablation of lithium manganate. *Appl. Surf. Sci.* **2007**, *254*, 811–815. [CrossRef]
38. Ershov-Pavlov, E.A.; Katsalap, K.Y.; Stepanov, K.L.; Stankevich, Y.A. Time-space distribution of laser-induced plasma parameters and its influence on emission spectra of the laser plumes. *Spectrochim. Acta Part B At. Spectrosc.* **2008**, *63*, 1024–1037. [CrossRef]
39. Kramida, A.; Ralchenko, Y.; Reader, J. NIST ASD Team, NIST Atomic Spectra Database Lines Form, NIST At. Spectra Database (Ver. 5.2). 2014. Available online: http://physics.nist.gov/asd (accessed on 9 October 2018).
40. Rao, K.H.; Smijesh, N.; Nivas, J.J.; Philip, R. Ultrafast laser produced zinc plasma: Stark broadening of emission lines in nitrogen ambient. *Phys. Plasmas* **2016**, *23*, 43503. [CrossRef]
41. Aguilera, J.A.; Aragón, C. Multi-element Saha-Boltzmann and Boltzmann plots in laser-induced plasmas. *Spectrochim. Acta-Part B At. Spectrosc.* **2007**, *62*, 378–385. [CrossRef]
42. Irimiciuc, S.A.; Gurlui, S.; Agop, M. Particle distribution in transient plasmas generated by ns-laser ablation on ternary metallic alloys. *Appl. Phys. B* **2019**, *125*, 190. [CrossRef]
43. Irimiciuc, S.A.; Bulai, G.; Gurlui, S.; Agop, M. On the separation of particle flow during pulse laser deposition of heterogeneous materials—A multi-fractal approach. *Powder Technol.* **2018**, *339*, 273–280. [CrossRef]
44. Lee, Y.I.; Sawan, S.P.; Thiem, T.L.; Teng, Y.Y.; Sneddon, J. Interaction of a laser beam with metals. Part II: Space-resolved studies of laser-ablated plasma emission. *Appl. Spec.* **1992**, *46*, 436–441. [CrossRef]
45. Milan, M.; Laserna, J.J. Diagnostics of silicon plasmas produced by visible nanosecond laser ablation. *Spectrochim. Acta-Part B At. Spectrosc.* **2001**, *56*, 275–288. [CrossRef]
46. Irimiciuc, S.A.; Gurlui, S.; Nica, P.; Focsa, C.; Agop, M. A compact non-differential approach for modeling laser ablation plasma dynamics. *J. Appl. Phys.* **2017**, *12*, 083301. [CrossRef]
47. Irimiciuc, S.A.; Agop, M.; Nica, P.; Gurlui, S.; Mihaileanu, D.; Toma, S.; Focsa, C. Dispersive effects in laser ablation plasmas. *Jpn. J. Appl. Phys.* **2014**, *53*, 116202. [CrossRef]
48. Merches, I.; Agop, M. *Differentiability and Fractality in Dynamics of Physical Systems*; World Scientific: Singapore, 2015.
49. Mandelbrot, B. *The Fractal Geometry of Nature*; WH Freeman Publisher: New York, NY, USA, 1993.
50. Nottale, L. *Scale Relativity and Fractal Space-Time: An Approach to Unifying Relativity and Quantum Mechanics*; Imperial College Press: London, UK, 2011.
51. Agop, M.; Merches, I. *Operational Procedures Describing Physical Systems*; CRC Press, Taylor and Francis Group: London, UK, 2019.
52. Mihaileanu, M. *Differential, Projective and Analytical Geometry*; Didactical and Pedagogical Publishing House: Bucuresti, Romania, 1972.

© 2019 by the authors. Licensee MDPI, Basel, Switzerland. This article is an open access article distributed under the terms and conditions of the Creative Commons Attribution (CC BY) license (http://creativecommons.org/licenses/by/4.0/).

Article

Investigations of Transient Plasma Generated by Laser Ablation of Hydroxyapatite during the Pulsed Laser Deposition Process

Maricel Agop [1,2], Nicanor Cimpoesu [3], Silviu Gurlui [4] and Stefan Andrei Irimiciuc [5,*]

1. Department of Physics, "Gh. Asachi" Technical University of Iasi, 700050 Iasi, Romania; m.agop@yahoo.com
2. Romanian Scientists Academy, 050094 Bucharest, Romania
3. Faculty of Materials Science and Engineering, "Gh. Asachi" Technical University of Iasi, 700050 Iasi, Romania; nicanornick@yahoo.com
4. Faculty of Physics, LOA-SL, Alexandru Ioan Cuza University of Iasi, 700506 Iasi, Romania; sgurlui@uaic.ro
5. Institute for Laser, Plasma and Radiation Physics, 077125 Bucharest, Romania
* Correspondence: stefan.irimiciuc@inflpr.ro

Received: 15 December 2019; Accepted: 7 January 2020; Published: 9 January 2020

Abstract: The optimization of the pulsed laser deposition process was attempted here for the generation of hydroxyapatite thin films. The deposition process was monitored with an ICCD (Intensified Coupled Charged Device) fast gated camera and a high-resolution spectrometer. The global dynamics of the laser produced plasma showed a self-structuring into three components with different composition and kinetics. The optical emission spectroscopy revealed the formation of a stoichiometric plasma and proved that the segregation in the kinetic energy of the plasma structure is also reflected by the individual energies of the ejected particles. Atomic Force Microscopy was also implemented to investigate the properties and the quality of the deposited film. The presence of micrometric clusters was seen at a high laser fluence deposition with in-situ ICCD imaging. We developed a fractal model based on Schrödinger type functionalities. The model can cover the distribution of the excited states in the laser produced plasma. Moreover, we proved that SL(2R) invariance can facilitate plasma substructures synchronization through a self-modulation in amplitude.

Keywords: pulsed laser deposition; plasma diagnostic; hydroxyapatite thin film; plasma structuring; SL(2R) invariance

---

1. Introduction

In the past few years, hydroxyapatite has drawn significant interest in the research community as well as the medical community for use in synthetic bones substitutes. Although hydroxyapatite (HA) is already used as part of fillers for treating various bone defects, their application can be restricted due to brittle fracturing of ceramic components [1]. Alternatively, due to their better physico-chemical, mechanical, and biological properties, titanium alloy or even titanium are widely used for load bearing applications such as bone plates, screws, and artificial joints. In literature there are reported implementations of a vast range of deposition techniques with the aim to obtain high quality hydroxyapatite coatings [2,3]. The stand out technique which gave the most promising results was proven to be Pulsed Laser Deposition (PLD) [3]. PLD is a multivariable technique which can offer the possibility of tailoring the composition, structure, and properties of the generated film simply by adjusting some of its control parameters (laser pulse geometry, laser fluence, wavelength or repetition frequency, target-substrate distance, substrate temperature, deposition time, etc.) [4]. The PLD technique has been advanced as an alternative to other more established and classical methods, and has an added advantage of enabling complete stoichiometric transfer from the target to the substrate.

PLD technique also possesses the ability to form desired film thickness, morphology, and composition by varying the deposition parameters. This method moreover offers a wide range of possibilities in terms of the nature of the irradiated targets and physio-chemical and biological properties single substrates for functional coatings. Some of the most outstanding results of hydroxyapatite (HA) coatings were reported for HA coated Ti, which reduced the fixation period to 3–6 months [3]. Although the advantages of PLD technique are well known, the tailoring of the technique can be time and resource consuming. An alternative approach is envisioned to understand the fundamental kinetics of the ejected particles and their influence over the properties of the thin films. This approach is based on implementation of in situ plasma characterization techniques. The diagnostics techniques have a proven record of showcasing various aspects of the laser produced plasmas with direct implications for pulsed laser deposition technology [5].

In this paper, we report the in-situ investigations of a transient plasma generated on a hydroxyapatite target during the pulsed laser deposition process. ICCD (Intensified Coupled Charged Device) fast camera imaging and optical emission spectroscopy are implemented to showcase different facets of the phenomena. The thin film deposited is investigated by using surface analysis techniques and reveals the best deposition conditions. A fractal model is developed based on Schrodinger type functionalities that can attest to various dynamics seen in HA laser produced plasma.

2. Experimental Set-Up

The experiments were performed using the installation presented in detail in reference [5]. Briefly, plasma was generated by irradiation of the hydroxyapatite target with a pulse generated by an Nd:YAG (10 ns, 10 Hz, 532 nm) laser, with an energy per pulse of 80 mJ and 40 mJ. The target used was a disk-shaped (1 cm in diameter) HA sample and was placed on an XYZ precision displacement system and constantly shifted during irradiation to provide new surfaces for each irradiation occurrence. All the experiments were performed at a background pressure of 10^{-2} Torr.

The samples of HA deposited by using PLD with thicknesses of 500 nm (P1) and 1000 nm (P2) were analyzed by using atomic force microscopy. The results were obtained on Park NX10 equipment. Two flexion scanning systems for XY and Z axes, respectively, were used that were independent for the sample and for the tip of the probe. These systems involved horizontal and orthogonal XY scanning with a reduced residual arc, movement outside the plane with less than 1 nm over the entire scan range, and the Z linearity deviation of the scanner being less than 0.015% over a full scan range. During the analysis of the surface morphology, experiments were performed using the Pin On module to determine the mechanical properties of the layers. The experiments were performed on the following equipment: Scanning Electron Microscopy (SEM) and Atomic Force Microscopy (AFM) for the analysis of the surface of the thin layers and Energy Dispersive X-Ray Spectroscopy (EDS) for chemical determinations. AFM analyses on EasyScan 2 equipment used a non-contact cantilever. Scans were performed on 12 and 50 µm squares, and the results were obtained in the forms of 2D and 3D, while surface roughness parameters were also provided. The images show a smooth surface with a roughness with nanometric values (around 1 nm), which in some areas show larger particles of material (these are in the order of micrometers according to SEM analyses).

3. Fast ICCD Camera Measurements

To track the global evolutions of the transient plasmas produced by laser ablation on the hydroxyapatite target, the technique of ICCD fast camera imaging was implemented. The role of these investigations was to determine the structure, to study the dynamic collective behavior of the ejected particles, and to understand the effect of the target on the global dynamics of the laser produced plasmas. Figure 1 shows snapshots of the laser produced plasma on a hydroxyapatite target recorded at various moments in time with respect to the laser-target impact moment. It can be observed that, during expansion, the plasma increases its volume and the position of the area with high emissivity (estimated as the center of mass) moves towards higher distances. In this work, the investigations were performed at low pressure and thus according to reference [6], this displacement was defined by a linear function, which translates as an expansion with a constant velocity. The slope of this linear representation defines the global expansion velocity of the plasma. With the increase of the background gas pressure there are reports [7] that show significant changes in the plasma dynamics: increased optical emission coupled with a confinement in the target region with an overall kinetic behavior being described in this case by a "drag" type function [8].

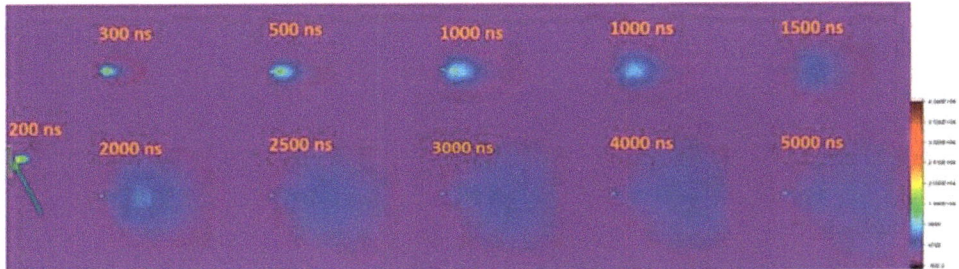

Figure 1. ICCD fast gated camera snapshots of laser produced plasmas on hydroxyapatite samples (gate width10 ns exposure and a laser fluence of 5 J/cm).

Figure 2 shows an image of the ablation plasma recorded after 350 ns from the interaction of the laser pulse radiation with the target and a cross section on the main expansion axis (where the separation in three structures is highlighted) and on a series of three axes parallel to the target surface (where the symmetry of the plasma with respect to the transverse axis is highlighted). The important feature highlighted by these images is the presence of a plasma separation process in three structures: the first structure (fast structure), the second structure (slow structure), and the existence of a third structure defined by small volume, high emission intensity, and low expansion velocity. The latter structure is usually known in the literature to describe the excitation of heavier and more complex objects ejected from the target such as clusters, molecules, or nanoparticles. This phenomenon was also observed for metals [9] or graphite [10] plasmas and was found numerically in a hydrodynamic fractal model [11]. Various reports from literature show that the observed structures move at constant speeds of the order of 10^4 m/s for the first structure, 10^3 m/s for the second structure, and hundreds of m/s for the thirds structure. Here, the experimental results show a good agreement between the obtained values and other reported in the literature [5–10]. The generated plasma is characterized by a global expansion speed of 4.5 km/s for the slow structure, 17 km/s for the fast structure, 300 m/s for the cluster structure.

Let us note that although the exact values are in line with others reported in the literature, it is worth underlining that these values need to be understood on a case to case scenario because the expansion velocities values strongly depend on the experimental conditions, nature of the material, and laser properties. Our group recently shown some dependencies between the expansion velocity and the atomic mass [9]. We also reported on the particle energy distribution in laser produced plasmas on complex targets [6,7] and on the heterogenic energy distribution in ablation plasmas.

The general assumption over the separation of laser produced plasmas into multiple components is that it comes as an effect of the interactions between the plasma particles and those of the background gas [10]. However, given the high vacuum conditions in which our experiments were performed, the plasma separation process may be related to the different ejection mechanisms that are seen in the case on nanosecond laser ablation. Thus, the fast structure is due to the presence of electrostatic mechanisms, such as the Coulomb explosion [9], while the slow structure is due to the presence of thermal mechanisms (such as phase explosions [12]). The results also show the presence of a third component of the plasma (identified in Figure 2 through zone 1). This is present in the vicinity of the target and has expansion speeds of 300 m/s. Figure 2 shows that the structure has a small angular distribution, compared to the classical structures, and is defined by a strong emission. These properties are characteristic, according to the results of the specific literature [13,14], and involve the dynamics of nanoparticle clusters or molecules that can be found in plasma in the Knudsen layer region.

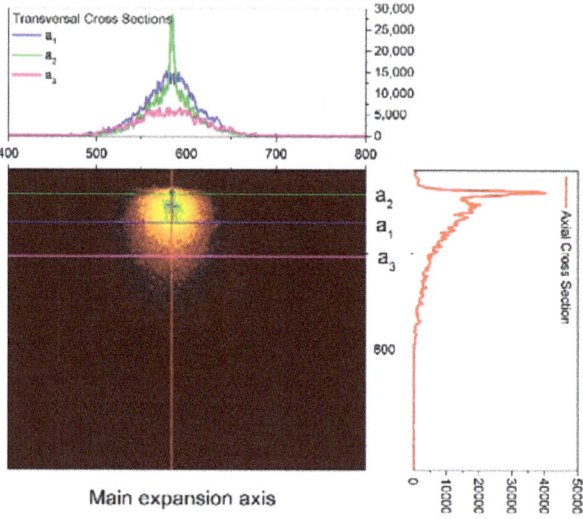

Figure 2. The image of a hydroxyapatite plasma recorded with the fast-gated camera after 300 ns and the corresponding cross sections.

4. Optical Emission Spectroscopy

The recording of the global emission by fast photography only gives a preliminary image of the dynamics of the plasma, and is not able to highlight the individual contributions of each species present in the plasma. For the purpose of separating these contributions, a temporal and spatially resolved spectral study was performed using the optical emission spectroscopy technique. In order to have an overview of the species present in the plasma, a global spectrum was recorded (using a relatively high integration time of ~1 μs). The results are shown in Figure 3. Based on the recorded spectra, the nature of the species present in the plasma can be identified using the authorized databases [15]. In the spectra, we have identified characteristic lines for the atoms and ions of the Ca species. The abundance of lines per each species is not stoichiometric, with the main lines corresponding to the Ca species. This is due to the difference in mass between Ca, O, and P, coupled with strong differences in collision frequency and discrepancies in the energetic levels of each ablated species [13]. This type of experiment manages to highlight very well the complex nature of hydroxyapatite, as shown in Figure 3. In this figure, it can be observed that the global emission spectrum has predominantly characteristic lines Ca, for which we have found correspondences for both atoms and ions.

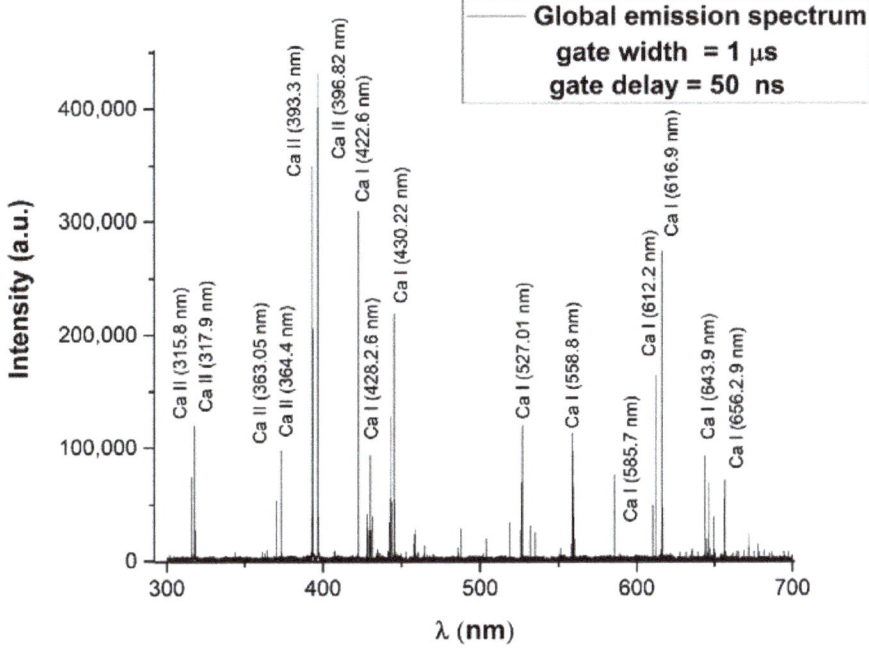

Figure 3. Global emission spectrum, acquired with a gate width of 1 μs at 50 ns across the main expansion axis.

To determine the excitation temperature of the species present in the plasma, we used the Boltzmann method [11,13]. In Figure 4 it can be seen that the representations for atoms and ions are described by linear decreases over a wide range of energies, which can be attributed to the energetic levels excited in the plasma. The presence of this type of dependence indicates the existence of a local thermodynamic equilibrium. Moreover, we opted to represent the ionic and atomic species separately in order to observe any heterogeneity in the energy of the ejected species. By analyzing the optical emission spectra and implementing the aforementioned Boltzmann plot method, we determined the excitation temperatures for Ca I (~0.3 eV), Ca II (~1.58 eV). Recent studies published by our group relating to laser ablation of metals show an inverse proportionality relationship between the global values of the excitation temperatures and the atomic mass of the elements [9]. Of course, one can comment on the validity of the global value of the excitation temperature in the context of the fast variation of plasma parameters. In literature [6,7,10] it was reported that the space time evolution of the temperature is defined by an increase for small distance and short evolution times followed by a decrease of the excitation temperatures specific to atoms and ions as the plasma expands. This behavior faithfully follows the global emission distribution of plasmas. An important aspect highlighted by our study is the differences found between the excitation temperatures of the atoms and ions in the same plasma. This result can be regarded as an effect of the differential heating by the ns laser pulse.

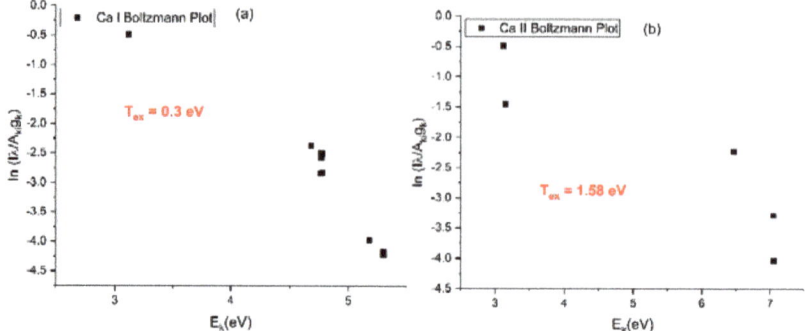

Figure 4. Boltzmann representation for Ca atoms (**a**) and ions (**b**) in HA-plasma. (where λ is the transition wavelength, A_{ki} is he Einstein coefficient of the k-i transition, g_k the statistical weight of the upper level).

In order to be able to differentiate between the individual contributions of each species, the space-time evolution of the emission lines characteristic to the plasma atoms and ions was followed. The results of these studies are shown in Figure 5, where the characteristic signals are shown for Ca I (396.15 nm) and Ca II (546 nm). The first emission spectral lines detected after the laser pulse are those of the ions (sign of a high expansion velocity). These are followed by the lines corresponding to the atoms. Such an analysis allows the investigation and separation of the dynamics of the ejected particles after determining their expansion velocity.

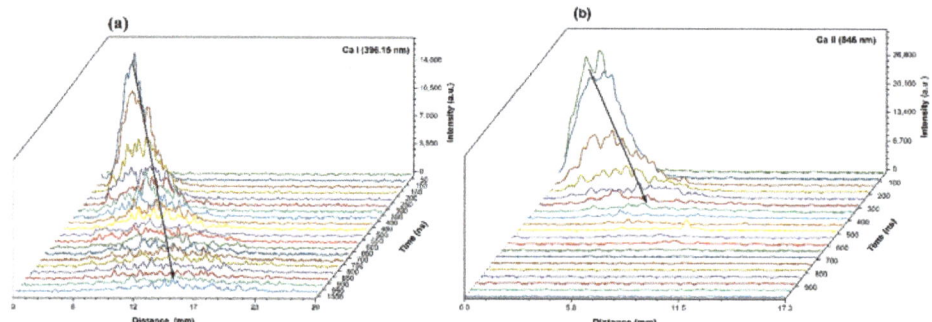

Figure 5. The comparative spatial-temporal evolution of the emission lines corresponding to Ca I atoms (396.15 nm)—(**a**) and Ca II ions (546 nm)—(**b**).

At the same time, we can see that the spatial profiles are structured into two main groups: a fast one, represented by the spectral lines of the ions, and a slower one, due mainly to the contribution of the neutrals. Such behavior may be related to the temporal evolution of the plasma recorded by fast photography with the ICCD camera. The velocity analysis of these groups confirms that the first structure observed by fast photography consists mainly of ionic species, while the second structure consists of neutral atoms.

In order to determine the expansion velocity of the individual species and then to compare with the global speeds presented above, the spatial evolutions of the intensities of the respective species were represented at different time points (Figure 6). It is observed that the emission maximum undergoes a displacement to greater distances with the evolution of the plasma. By representing this spatial-temporal variation and fitting with a linear function (Figure 6b), the expansion velocities of atoms (Ca I—8.7 km/s) and ions (Ca II—16.3 km/s) were determined. The evolution of plasma species velocities highlights a link between the degree of ionization and the expansion velocity. This dependence specifies the influence of the fundamental particle ejection mechanisms. In general, atomic species are removed by thermal mechanisms that impose an inverse proportional time link between the velocity and the square root of the atomic mass. Plasma ions do not have the same type of dependence, most of them being removed by electrostatic mechanisms, which can account for the difference in expansion velocity between the two species.

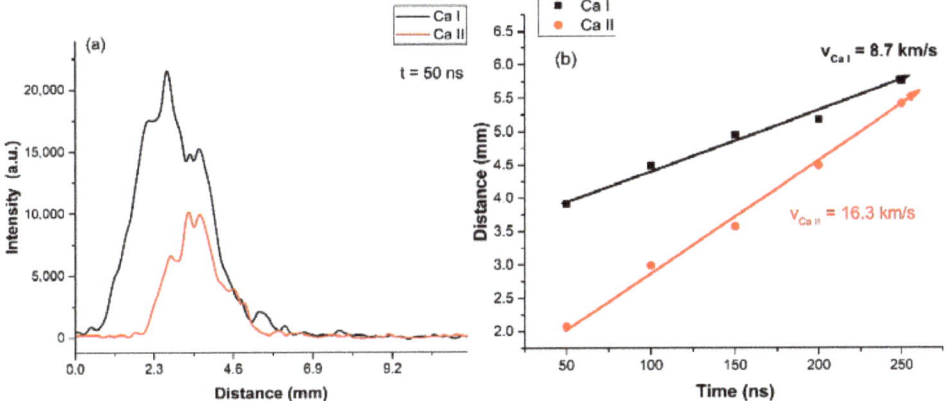

Figure 6. The comparative spatial distribution of Ca atoms and ions recorded after 50 ns (**a**) and the dependence of the maximum emission in space-time for the two species (**b**).

The heterogeneity of the ablated ions and atoms could seen from an energetic perspective, with significant kinetic and thermal energies recorded for the Ca atoms and ions coupled with the structuring in three plasma structure with distinct inner composition. All these factors would effectively affect both the quality and the stoichiometry of the HA film. In the following section (Sections 5 and 6), we tested this statement by employing surface analysis techniques to characterize the films generated in the conditions discussed in the Sections 3 and 4.

5. Surface Investigations of Hydroxyapatite Thin Films

Figure 7 shows the results obtained by scanning the surfaces of the two samples: P1 (500 nm) and P2 (1000 nm). Sample P1 was obtained by depositing with a fluency of 2.5 J/cm^2, and sample P2 with a fluency of 5 J/cm^2, for 10 min. 2D micrographs were made on a surface of 50 × 50 µm^2, see Figure 7a for sample P1 and Figure 7c for sample P2. In both scans, a fine surface of the realized layer is observed in a zone covered by droplets of material, with a higher density of drops in the case of sample P2.

The 2D images (Figure 7b,d) show the state of the surface of the thin layer and the distribution of the larger particles of distant material on the surface of the thin layer. From what is observed, sample P2 shows more droplets of material on the surface with larger dimensions and a higher roughness (and a value of 30 nm) [16]. These finding are in line with the ICCD fast camera imaging results, where we noticed the presence of a third plasma structure containing clusters and nanoparticles. Even after reducing the fluence, we still noticed a small density of clusters on the films, meaning that the lower density of the sample induces naturally clusters in the laser produced plasmas.

SEM analyses (Figure 8) were performed by using SEM VegaTescan LMH II equipment, as well as a secondary electron (SE) detector and a 30-kV electron cannon supply voltage. The thin films were analyzed in high vacuum using a layered support on carbon-band glass support. In the images, the formation of a layer with a good structural homogeneity and with the appearance of larger particles of material can be observed [17].

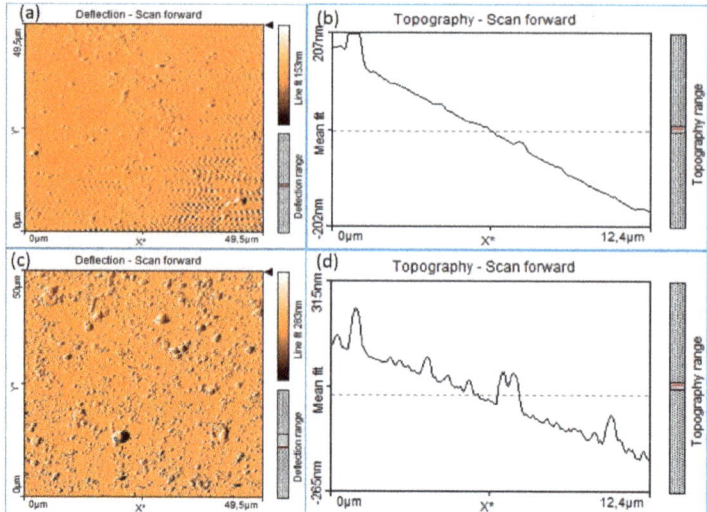

Figure 7. AFM images of HA thin layers deposited by PLD (**a**,**b**) for the sample surface P1 and (**c**,**d**) for the sample surface P2.

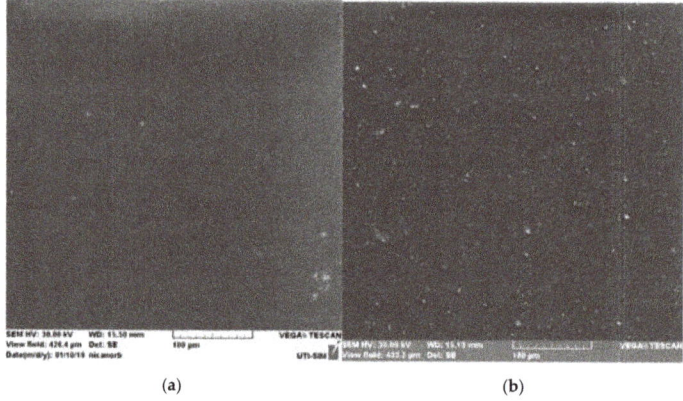

Figure 8. SEM images of the surface of surface layers of HA deposited by PLD for sample P1 (**a**) and for sample P2 (**b**).

For sample P2, the droplets of material, generally round in shape, show dimensions from 100 nm to 3 μm. On an area of 200 μm^2, an average number of 35 particles, of which only one had a diameter greater than 2 μm, was identified. On sample P1, there were much less particles on the homogeneous surface of the layer, with dimensions also between 100 nm and 2 μm.

EDS analyses were performed on Bruker equipment using the Element List analysis mode. The analysis mainly followed the evolution of the Ca: P ratio obtained on the thin layer compared to the HA target used in the ablation process, with the value of 1.67 [3]. The elements identified on the HA layers, were O, Ca, and P. Table 1 shows the results of the chemical analysis on the two thin films of HA obtained by laser ablation. The results were obtained by mediating 10 chemical compositions taken from 10 different surfaces. Due to the very small dimensions of the layers obtained (<1 μm), the main element obtained is the oxygen that is also part of the target compound HA. The ratios obtained between Ca and P are very good compared to the ratio between them in the HA target. We can thus say that, for a bio-compatible material such as HA, the stoichiometric transfer takes place at low fluences

(<2.5 J/cm^2), a result confirmed by similar values found in the literature [18]. For high fluency, as mentioned above, the layer has high roughness and has nano- and micro-metric structures. In the case of the higher fluency, a deviation from the target stoichiometry is observed. Based on this in order to obtain stoichiometric transfer on the deposition of thin layers of HA through laser ablation, it is desirable to use a relatively small fluence.

Table 1. Composition of the component elements of the HA thin layers obtained by laser ablation, for P1 and P2.

Sample	O		Ca		P		Ratio Ca:P
	wt.%	at%	wt.%	at%	wt.%	at%	
P1	97.87	98.87	1.32	0.71	0.78	0.42	1.69
P2	97.29	98.56	1.73	0.92	0.98	0.52	1.77
EDS error (%)	1.5		0.2		0.1		

6. Mechanical Properties of Thin Layers

6.1. Analysis of the 500 nm HA Layer (P1)

Analysis of the morphology of the deposited layer showed that it is composed of a basic homogeneous layer, but also has some larger formations of materials such as hydroxyapatite (which helps to anchor the layer in contact with the biological material). The formations have heights of up to 100 nm and the deposited layer has surface fluctuations of up to 20 nm.

The mechanical properties (adhesion [nN], deformability [nm], and modulus of elasticity [GPa]) are shown in Figure 9 on the selected surface. The homogeneity of the distribution of adhesion and of the modulus of elasticity shows that a layer with good properties and characteristics close to those of the massive material was obtained. Hydroxyapatite is a ceramic material whose plasticity is recognized [16,19], but in the form of a thin layer, it can borrow from the characteristics of the substrate.

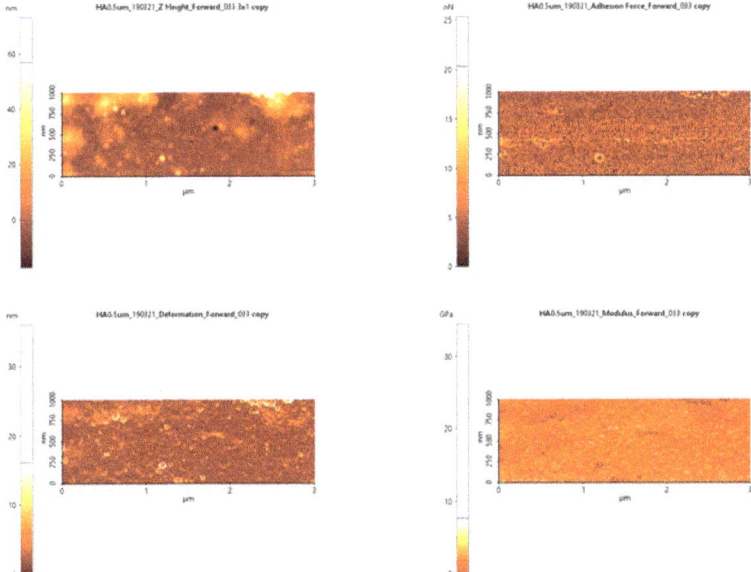

Figure 9. Analysis of the mechanical properties of the 500 nm HA layer (P1).

6.2. Analysis of the 1000 nm HA Layer (P2)

In the case of the larger 1000 nm layer, here we can find a number of larger HA particles. These can be considered as a separate material because they have different dimensions from the base layer. In height they do not exceed 150 nanometers but are large enough (1–2 µm in diameter) to influence the characteristics of the base layer.

Figure 10 shows an analysis of the mechanical properties of the HA layer (1000 nm). The particles of material increased, rising from 200 nm in height. The adhesion force increased compared to the 500 nm layer shown in Figure 10, reaching values of 40 nN. The drops of HA material do not have an additional adhesive force because they are less attached to the substrate.

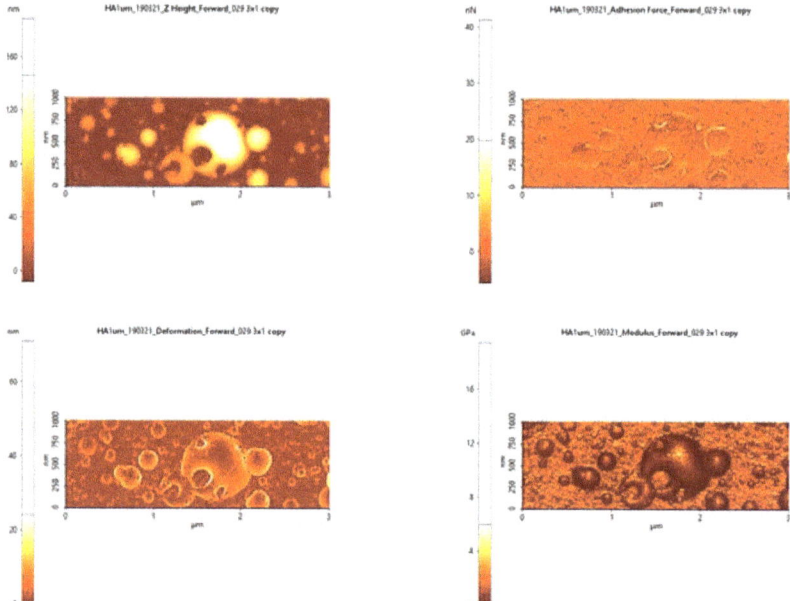

Figure 10. Analysis of the mechanical properties of the 1000 nm HA layer (P2).

The drops of material that were formed after the base layer had a lower modulus of elasticity (see Figure 10) compared to the base layer, which led to a differentiation of the properties of the ceramic material. The susceptibility to deformation was higher for the droplet material and at higher values for the material with a thickness of 1000 nm, compared to materials with a thickness of 500 nm.

7. Theoretical Model

The merit for the use of a complex theoretical model that can cover a wide range of behavior of transient plasmas has been shown by our group in the past few years in a series of papers that cover: dissipative or dispersive phenomena, particle heterogeneity in laser produced plasma on metallic alloys, oscillatory behavior of charged particles, plasma self-structuring, and stoichiometry issues during the pulsed laser deposition process. The fractal paradigm was proven to be a proper media for the development viable theoretical models used to describe laser produced plasmas complex dynamics. In the following section (Section 7), we focus on studying the dynamics of the laser ablation plasma in a nondifferentiable regime of the Schrodinger type at various resolution scales.

Let us first consider the movement equation (geodesics equation) of the ablation plasma structural units [20]:

$$\frac{d\hat{V}^l}{dt} = \partial_t \hat{V}^l + \hat{V}^i \partial_i \hat{V}^l - \chi (dt)^{(\frac{2}{D_F})-1} \partial_i \partial^i \hat{V}^i \tag{1}$$

where

$$\partial_t = \frac{\partial}{\partial t}, \partial_l = \frac{\partial}{\partial x^l}, \partial_l \partial_l = \frac{\partial}{\partial x^l}\frac{\partial}{\partial x^l} \quad (2)$$

$$\hat{V}^i = V^i - iU^i, \, i = \sqrt{-1}. \quad (3)$$

In relations (1)–(3), x^l is the fractal spatial coordinate, t is the nonfractal temporal coordinate with the role of an affine parameter of the movement curve, $\hat{V}^i$ is the complex velocity, V^i is the scale resolution independent real part of $\hat{V}^i$, U^l is the scale resolution (dt) dependent imaginary part of $\hat{V}^i$, λ is the "diffusion coefficient" associated with the fractal nonfractal transition, and D_F is the movement curve fractal dimension. For D_F we can choose any accepted definition: the Kolmogorov definition, the Haussdorff-Besikovici definition, etc. [21]. However once one definition is accepted, it has to be constant.

Let us consider that the particle ejection as a result of the laser-target interaction is defined through an irrotational movement of the ablation plasma structural units. Then:

$$\hat{V}^i = -2i\lambda (dt)^{(2/D_F)-1} \partial^i \ln \Psi \quad (4)$$

where Ψ is the fractal state function, while $\ln\Psi$ is the scalar potential of the complex velocity field. In these conditions by substituting (3) in (1) and following the procedure from [20], we obtain the Schrödinger equation of fractal type (free of any constraints):

$$\lambda^2 (dt)^{(\frac{4}{D_F})-2} \partial^l \partial_l \Psi + i\lambda (dt)^{(\frac{2}{D_F})-1} \partial_t \Psi = 0. \quad (5)$$

In the case of external constraints given by the scalar potential U, the Schrödinger equation of fractal type becomes:

$$\lambda^2 (dt)^{(\frac{4}{D_F})-2} \partial^l \partial_l \Psi + i\lambda (dt)^{(\frac{2}{D_F})-1} \partial_t \Psi - \frac{U}{2}\Psi = 0. \quad (6)$$

If we consider a simplified version of this approach by considering the stationary one-dimensional case (i.e., for $\Psi(x,t) \to \chi(x)$ through the method from reference [20]), restricted by a potential barrier represented in Figure 11, we can find:

$$\frac{\partial^2 \chi}{\partial x^2} + \frac{(V_0 - E)}{2m_0 \lambda^2 (dt)^{(4/D_F)-2}} \chi \quad (7)$$

where $\chi = \chi(x)$ is the fractal stationary state function, E_0 is the fractal energy of the ablation plasma structural unit [22] and the m_0 is the rest mass of the same structural unit.

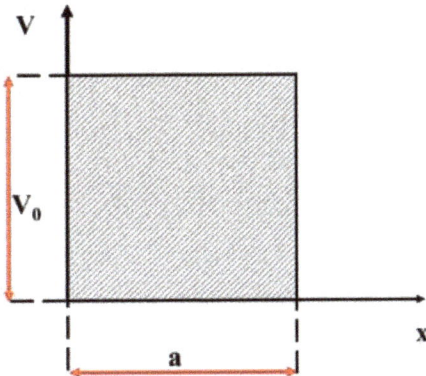

Figure 11. External scalar potential for the fractal type tunneling effect.

The potential barrier in this case will differentiate between two possible states of the ejected particles in during the expansion (emissive and non-emissive). In this way, we will be able to discriminate between the particles that are recorded using the investigation techniques discussed in the previous sections (Sections 3 and 4) and those who are not measured explicitly. In such a context, by implementing the method from reference [23], we can find the two coefficients that can be associated with the optical emission process, one describing the excited states of a family of structural units described by the same fractalization degree (I), while the second coefficient defines a different family of states describing the non-excited states of the plasma (N):

$$N = \frac{\left(X^2 + Y^2\right)^2}{\left(Y^2 - X^2\right)^2 + 4Y^2 X^2 \coth^2(Y)} \qquad (8)$$

$$I = \frac{4X^2 Y^2}{4X^2 Y^2 + \left(X^2 + Y^2\right)^2 \sinh^2(Y)} \qquad (9)$$

with

$$X = \left[\frac{E}{2m_0 \lambda^2 (dt)^{(4/D_F)-2}}\right]^{1/2} a \qquad (10)$$

$$Y = \left[\frac{V_0 - E}{2m_0 \lambda^2 (dt)^{(4/D_F)-2}}\right]^{1/2} a. \qquad (11)$$

In Figure 12 we have represented the evolution of the I coefficient for different X-Y fractal coordinates. The Y parameters were used to specify the fractalization of the simulated structural units. We noticed that for various values of the Y (fractalization values attributed to different species in the laser produced plasma), distribution maxima were shifted to higher values of X. This particular behavior was showcased through ICCD fast camera imaging where we saw the global emission spatial distribution presenting multiple maxima characterizing different plasma structures (Figure 2) which were rich in one specific type of particle species (fast structure-ions, slow structure-atoms, last structure-clusters). Each structure, containing different particles, in the fractal paradigm will automatically be defined by a different fractalization degree.

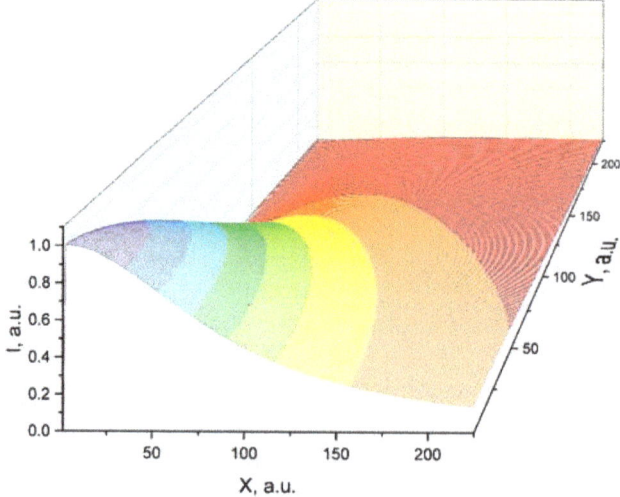

Figure 12. Modeling of the distribution of the excited states species of the laser produced plasma in the framework of a fractal paradigm.

Therefore, in order to accurately describe the plasma expansion in this representation, we need to account for the functionality of the scale superposition principle. This means that the dynamics of any complex system (here—laser produced plasmas) at a global scale resolution can be identified with the cumulative action of the dynamics at various local scale resolutions. In such a context, we considered the spectral emission representation (*I*) in a fractal representation for various scale resolutions and fractalization degrees (Figure 13). As a result, we could reconstruct the global distribution by convoluting the three distributions. The cumulative action also implied the interaction between the plasma substructures. Since Equation (7) is invariant with respect to the SL(2R) group [24], the interactions between the plasma substructures (Coulomb substructure, Thermal substructure, and Cluster substructure) were mathematically defined through the Stoka type procedure [25,26]. These procedures specify synchronizations amongst the structure though self-modulation in amplitude according to the results from reference [27]. These phenomena imply Stoler type transformations from a mathematical point of view [28]. These transformations [27,28] led to synchronization between the substructures of the laser ablation plasmas being achieved through charge creation and annihilation mechanisms.

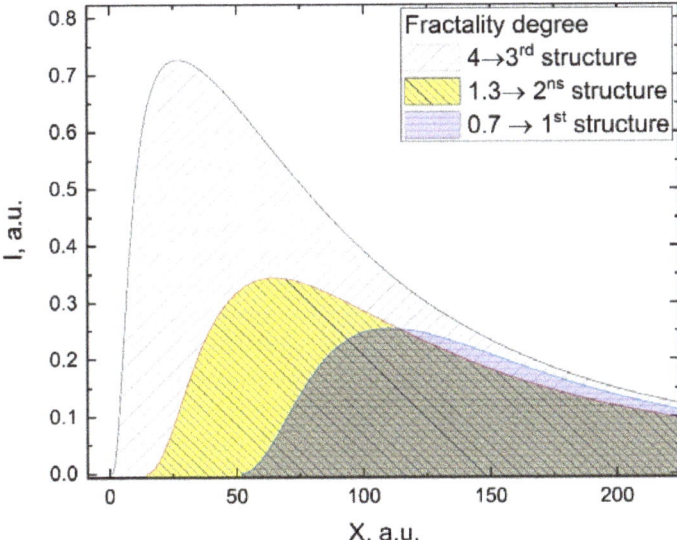

Figure 13. Selection of laser produced plasma entities spectral emission based on various fractalization degrees.

8. Conclusions

The process of deposition of thin films by laser ablation of hydroxyapatite was studied. Two complementary techniques were implemented: fast camera photography, and both spatially and temporally resolved emission spectroscopy. The fast camera photography allowed the identification of the plasma structures expanding with different velocities and having different geometries. Optical emission spectroscopy allowed the determination of the excitation temperatures of each component of the plasma (Ca, P, and O) and revealed a heterogeneity in the distribution of the internal energy of the plasma. Spatial-temporal evolution of the emission of Ca atoms and ions was also tracked and the expansion velocities of each species were determined.

Two thin layers were deposited under low and high fluency conditions. Complementary surface investigation techniques (SEM, AFM, and EDX) were implemented in order to find the right route for obtaining a thin film with complete stoichiometric transfer. For the high fluency, films with nano- and micrometric-sized structures and a small deviation from the stoichiometry due to these structures were obtained. Therefore the optimum conditions for smoother stoichiometric films were found to be at a fluence of 2.5 J/cm^2. We developed a fractal model based on Schrödinger type functionalities. The model can cover the distribution of the excited states in the laser produced plasma. Moreover, we proved that SL(2R) invariance can enable plasma substructure synchronization through self-modulation.

Author Contributions: All authors have read and agreed to the published version of the manuscript. Conceptualization, S.G., M.A. and S.A.I.; methodology S.G., N.C. and M.A.; investigation, S.A.I., N.C. and M.A.; writing—original draft preparation, S.A.I., M.A. and S.G.; writing—review and editing, S.A.I., N.C. and M.A.; visualization, S.A.I.; supervision, M.A. and S.G.

Funding: This work was financially supported by R National Authority for Scientific Research and Innovation in the framework of Nucleus Program–16N/2019.

Conflicts of Interest: The authors declare no conflict of interest.

References

1. Koch, C.F.; Johnson, S.; Kumar, D.; Jelinek, M.; Chrisey, D.B.; Doraiswamy, A.; Jin, C.; Narayan, R.J.; Mihailescu, I.N. Pulsed laser deposition of hydroxyapatite thin films. *Mater. Sci. Eng.* **2007**, *27*, 484–494. [CrossRef]
2. Fernández-Pradas, J.M.; Clèries, L.; Sardin, G.; Morenza, J.L. Hydroxyapatite coatings grown by pulsed laser deposition with a beam of 355 nm wavelength. *J. Mater. Res.* **1999**, *14*, 4715–4719. [CrossRef]
3. Rajesh, P.; Muraleedharan, C.V.; Komath, M.; Varma, H. Pulsed laser deposition of hydroxyapatite on titanium substrate with titania interlayer. *J. Mater. Sci. Mater. Med.* **2011**, *22*, 497–505. [CrossRef]
4. Bulai, G.; Trandafir, V.; Irimiciuc, S.A.; Ursu, L.; Focsa, C.; Gurlui, S. Influence of rare earth addition in cobalt ferrite thin films obtained by pulsed laser deposition. *Ceram. Int.* **2019**, *45*, 20165–20171. [CrossRef]
5. Irimiciuc, S.; Bulai, G.; Agop, M.; Gurlui, S. Influence of laser-produced plasma parameters on the deposition process: In situ space- and time-resolved optical emission spectroscopy and fractal modeling approach. *Appl. Phys. A Mater. Sci. Process.* **2018**, *124*, 1–14. [CrossRef]
6. Irimiciuc, S.; Boidin, R.; Bulai, G.; Gurlui, S.; Nemec, P.; Nazabal, V.; Focsa, C. Laser ablation of (GeSe2)100−x (Sb2Se3)x chalcogenide glasses: Influence of the target composition on the plasma plume dynamics. *Appl. Surf. Sci.* **2017**, *418*, 594–600. [CrossRef]
7. Canulescu, S.; Papadopoulou, E.L.; Anglos, D.; Lippert, T.; Schneider, C.W.; Wokaun, A. Mechanisms of the laser plume expansion during the ablation of LiMn2O4. *J. Appl. Phys.* **2009**, *105*, 063107. [CrossRef]
8. Aké, C.S.; De Castro, R.S.; Sobral, H.; Villagrán-Muniz, M. Plume dynamics of cross-beam pulsed laser ablation of graphite. *J. Appl. Phys.* **2006**, *100*, 053305.
9. Irimiciuc, S.A.; Nica, P.E.; Agop, M.; Focsa, C. Target properties–plasma dynamics relationship in laser ablation of metals: Common trends for fs, ps and ns irradiation regimes. *Appl. Surf. Sci.* **2019**, *506*, 144926. [CrossRef]
10. Harilal, S.S.; Bindhu, C.V.; Tillack, M.S.; Najmabadi, F.; Gaeris, A.C. Internal structure and expansion dynamics of laser ablation plumes into ambient gases. *J. Appl. Phys.* **2003**, *93*, 2380–2388. [CrossRef]
11. Irimiciuc, S.A.; Gurlui, S.; Nica, P.; Focsa, C.; Agop, M. A compact non-differential approach for modeling laser ablation plasma dynamics. *J. Appl. Phys.* **2017**, *12*, 083301. [CrossRef]
12. Harilal, S.S.; Bindhu, C.V.; Tillack, M.S.; Najmabadi, F.; Gaeris, A.C. Plume splitting and sharpening in laser-produced aluminium plasma. *J. Phys. D Appl. Phys.* **2002**, *35*, 2935–2938. [CrossRef]
13. Irimiciuc, S.A.; Gurlui, S.; Agop, M. Particle distribution in transient plasmas generated by ns-laser ablation on ternary metallic alloys. *Appl. Phys. B* **2019**, *125*, 190. [CrossRef]
14. Milan, M.; Laserna, J.J. Diagnostics of silicon plasmas produced by visible nanosecond laser ablation. *Spectrochim. Acta Part B At. Spectrosc.* **2001**, *56*, 275–288. [CrossRef]
15. Kramida, A.; Ralchenko, Y.; Reader, J. NIST ASD Team, NIST Atomic Spectra Database Lines Form, NIST At. Spectra Database (Ver. 5.2). 2014. Available online: http://physics.nist.gov/asd (accessed on 9 October 2018).
16. Song, J.; Liu, Y.; Zhang, Y.; Jiao, L. Mechanical properties of hydroxyapatite ceramics sintered form powders with different morphologies. *Mater. Sci. Eng. A* **2011**, *528*, 5421–5427. [CrossRef]
17. Wanga, D.G.; Chena, C.Z.; Yang, X.X.; Ming, X.C.; Zhang, W.L. Effect of bioglass addition on the properties of HA/BG composite films fabricated by pulsed laser deposition. *Ceram. Int.* **2018**, *44*, 14528–14533. [CrossRef]

18. Nishikawa, H.; Umatani, S. Effect of ablation laser pulse repetition rate on the surface protrusion density of hydroxyapatite thin films deposited using pulsed laser deposition. *Mater. Lett.* **2017**, *09*, 330–333. [CrossRef]
19. Cho, M.-Y.; Lee, D.-W.; Kim, I.-S.; Kim, W.-J.; Koo, S.-M.; Lee, D.; Kim, Y.-H.; Oh, J.-M. Evaluation of structural and mechanical properties of aerosol-deposited bioceramic films for orthodontic brackets. *Ceram. Int.* **2019**, *l45*, 6702–6711. [CrossRef]
20. Merches, I.; Agop, M. *Differentiability and Fractality in Dynamics of Physical Systems*; World Scientific: Singapore, 2016.
21. Mandelbrot, B. *The Fractal Geometry of Nature*; WH Freeman Publisher: New York, NY, USA, 1993.
22. Enescu, F.; Irimiciuc, S.A.; Cimpoesu, N.; Bedelean, H.; Bulai, G.; Gurlui, S.; Agop, M. Investigations of Laser Produced Plasmas Generated by Laser Ablation on Geomaterials. Experimental and Theoretical Aspects. *Symmetry* **2019**, *11*, 1391. [CrossRef]
23. Bujoreanu, C.; Irimiciuc, S.A.; Benchea, M.; Nedeff, F.; Agop, M. A fractal approach of the sound absorption behaviour of materials. Theoretical and experimental aspects. *Int. J. Non-Linear Mech.* **2018**, *103*, 128–137. [CrossRef]
24. Mihaileanu, M. *Differential, Projective and Analytical Geometry*; Didactical and Pedagogical Publishing House: Bucuresti, Romania, 1972.
25. Stoka, M.I. *Integral Geometry*; Romanian Academy Publishing House: Bucharest, Romania, 1967.
26. Mazilu, N.; Agop, M. *Skyrmions—A Great Finishing Touch to a Classical Newtonian Phylosiphy*; Nova Publishing: New York, NY, USA, 2012.
27. Irimiciuc, S.A.; Bulai, G.; Gurlui, S.; Agop, M. On the separation of particle flow during pulse laser deposition of heterogeneous materials—A multi-fractal approach. *Powder Technol.* **2018**, *339*, 273–280. [CrossRef]
28. Stoler, D. Equivalence Classes of Minimum Uncertainty Packets. *Phys. Rev. D* **1970**, *1*, 3217. [CrossRef]

© 2020 by the authors. Licensee MDPI, Basel, Switzerland. This article is an open access article distributed under the terms and conditions of the Creative Commons Attribution (CC BY) license (http://creativecommons.org/licenses/by/4.0/).

Article

In-Situ Plasma Monitoring during the Pulsed Laser Deposition of $Ni_{60}Ti_{40}$ Thin Films

Nicanor Cimpoesu [1], Silviu Gurlui [2], Georgiana Bulai [3], Ramona Cimpoesu [1,*], Viorel-Puiu Paun [4], Stefan Andrei Irimiciuc [5,*] and Maricel Agop [6,7]

1. Materials Science and Engineering Faculty, Technical University "Gh. Asachi" from Iasi, 700506 Iași, Romania; nicanor.cimpoesu@tuiasi.ro
2. Physic Faculty, "Al. I. Cuza" University from Iasi, 700506 Iași, Romania; sgurlui@uaic.ro
3. Integrated Center for Studies in Environmental Science for North-East Region (CERNESIM), Alexandru Ioan Cuza University of Iasi, 700506 Iasi, Romania; georgiana.bulai@uaic.ro
4. Physics Department, Faculty of Applied Sciences, University Politehnica of Bucharest, 010614 Bucharest, Romania; paun@physics.pub.ro
5. Department of Physics, "Gh. Asachi" Technical University of Iasi, 700050 Iasi, Romania
6. Romanian Scientists Academy, 54 Splaiul Independentei, 050094 Bucharest, Romania; m.agop@tuiasi.ro
7. National Institute for Laser, Plasma and Radiation Physics, 409 Atomistilor Street, 077125 Bucharest, Romania
* Correspondence: ramona.cimpoesu@tuiasi.ro (R.C.); stefan.irimciuc@inflpr.ro (S.A.I.)

Received: 14 December 2019; Accepted: 30 December 2019; Published: 6 January 2020

Abstract: The properties of pulsed laser deposited of $Ni_{60}Ti_{40}$ shape memory thin films generated in various deposition conditions were investigated. In-situ plasma monitoring was implemented by means of space- and time-resolved optical emission spectroscopy, and ICCD fast camera imaging. Structural and chemical analyses were performed on the thin films using SEM, AFM, EDS, and XRD equipment. The deposition parameters influence on the chemical composition of the thin films was investigated. The peeled layer presented on DSC a solid-state transformation in a different transformation domain compared to the target properties. A fractal model was used to describe the dynamics of laser produced plasma through various non-differentiable functionalities. Through hydrodynamic type regimes, space-time homographic transformations were correlated with the global dynamics of the ablation plasmas. Spatial simultaneity of homographic transformation through a special SL(2R) invariance implies the description of plasma dynamics through Riccati type equations, establishing correlations with the optical emission spectroscopy measurements.

Keywords: nitinol; pulsed laser deposition; in situ plasma monitoring; thin films; fractal modelling; SL(2R) invariance; homographic transformations; Riccati equation

1. Introduction

Due to the demand of applications in the field of engineering, the new materials are kept in a developing state in order to improve their performance, but also for creating new functions. Among these, there is a group of materials that are able to respond to a special stimulus through the alteration of the encircling physical and chemical features. These stimuli include the temperature (thermo-receptive materials), tension or pressure (mechanical-receptive materials), current or electrical tension (electro-receptive materials), magnetic field (magneto-receptive materials), change of the pH, of the solvent, or humidity (chemical-receptive materials) and light (photo-sensitive materials) [1–3].

Such a special class of materials are the shape memory alloys (SMA) that possess a range of desired properties, especially a high mechanical capacity in comparison with their weight, through which is developed the ability of recovery, a deformation produced by a massive transformation and a deformation due to heating and cooling, pseudo-elasticity (super elasticity), high capacity of damping, high chemical corrosion resistance, and bio-compatibility (in the case of alloys based on Ti and Fe) [4–6].

In the past few years shape memory alloy thin films were proven promising with high performance in the field of applications of micro-electric mechanical systems (M.E.M.S) because they can be implemented through standard techniques of lithography or depositing, fabricated at an industrial scale [7]. The thin layers of shape memory alloys need only a small amount of thermic capacity to heat or cool, which is why the answer time is substantially reduced and the working speed increases considerably. The effective power capacity on a volume corresponding to a layer obtained from an alloy with shape memory exceeds the one of the classic mechanisms proposed for micro-activation [8–10]. The transformation process of a bulk alloy with shape memory in thin film is accompanied by numerous important changes at the level of the mechanical, physical, chemical, electric, and optical properties such as the load resistance, elastic module, hardness, and attenuation capacity, regaining the initial form, electric resistivity, thermic conductivity, thermic expansion coefficient, surface rugosity, vapors permeability, and the dielectric constant, etc.

This transformation can be used in designing and fabricating micro-sensors and micro-actuators. However, due to the lack of a complete understanding of the AMF layers, combined with the difficulty of controlling the depositing parameters, these films have not benefitted of too much attention in the MEMS technology of production, in comparison to other technologies of obtaining the micro-actuators. Thin films based on Ni-Ti are the most SMA used materials under thin film form, and they are obtained through the atomization method. There are also depositing methods: laser ablation, depositing with ions fascicle, plating with jet of ion plasma, the plasma atomization and the blitz evaporation, but with certain inherent problems such as the lack of uniformity of the layer thickness and the composition's lack of uniformity, low coating speed, or low handling and incompatibility with the process performed by MEMS, etc. The depositing processes through laser ablation surpass the majority of these problems [11–14]. The transformation temperatures, behavior of shape memory, and the super-elasticity of the pulverized layers of Ni-Ti are sensitive at the most metallurgical factors (the alloy composition, contamination, thermic-mechanical treatments of re-annealing and ageing), at the atomization conditions (co-atomization with multi-aims, the target, the gas pressure, the distance from the target to the sub-layer, the depositing temperature, obliquity sub-layer, etc.), as well as to the application conditions (loading conditions, temperature of the environment, heat dissipation, heating/cooling speed, mechanical loading speed) [15,16]. The shape memory effect involves the crystallographic transformation thermo-elastic with reversible stage or the martensitic transformation from the initial stage of high temperature to that final stage of low temperature [17–22]. Generally speaking, there are 2 types of martensitic transformation, one with a single stage A→M and the other with 2 stages A→R→M, where A is the austenitic phase of high temperature, R is the rhombohedral intermediary phase, and M is the martensitic phase of low temperature. The experimental results presented an in-depth presentation of the efficiency of the pulsed laser deposition process towards obtaining shame memory alloy thin films and the effect of the deposition conditions on the properties of the thin film.

In this paper we report the experimental results on the deposition of thin films with special properties using Nitinol targets (shape memory alloy). Also, there are followed aspects about with the target behavior, and the transmitted plasma during the depositing and created thin layers. A fractal model used to describe the dynamics of laser produced plasma through various functionalities of non-differentiable type was established. Hydrodynamic type regimes by means of space time homographic transformations were corelated with the global dynamics of the ablation plasmas. In such a context, spatial simultaneity of homographic transformation through a special SL(2R) invariance implies the description of plasma dynamics through Riccati type equations can be used to describe the dynamics of ablation plasma seen through the optical emission spectroscopy measurements.

2. Experimental Set-Up

The properties of the thin films and the plasma investigation were performed using an Nd: YAG laser (λ = 266, 355 or 532 nm, pulse width 5 ns) working at a 10 Hz repetition rate. The laser beam was focused with lens with a focal distance f = 35 cm, on a target NiTi placed in a vacuum chamber ($p = 10^{-2}$ Torr). The NiTi alloy (nitinol60) was acquired from the firm Saes Getters [17], in softened state. The estimated diameter in the impact point was almost 400 µm. The energy laser fascicle was monitored continuously using a joule meter Ophir. The energy used usually was of 30 mJ/impulse, that leads to a typical intensity of the laser of 6 GW/cm^2.

The target of shape memory alloy is moved in the XY plan with a micrometer manipulator in order to expose a new area towards the laser fascicle (to obtain a uniform ablation, the target was rotated and translated simultaneously). The formation and dynamics of plume were studied with an intensified camera ICCD (PI MAX, 576_384, with minimum gate of 2 ns) placed perpendicularly on the direction of expansion of plasma, Figure 1a.

The spectral composition of plasma formed in the ablation process depends on the chemical composition of the sample, which was analyzed with a high resolution spectrometer. This technique reaches a high potential of using as analysis method in real time of the chemical elements.

In the case of analysis of the metallic ablated materials it was used an identification of the base spectral lines through the analogy with those identified in the NIST library [18]. The spatial–temporal dynamics of the transitory plasmas generated through laser ablation of high fluency with pulses at nanoseconds was analyzed using optical methods (optical emission spectroscopy and ICCD fast camera imaging), the analysis figure being presented in Figure 1b.

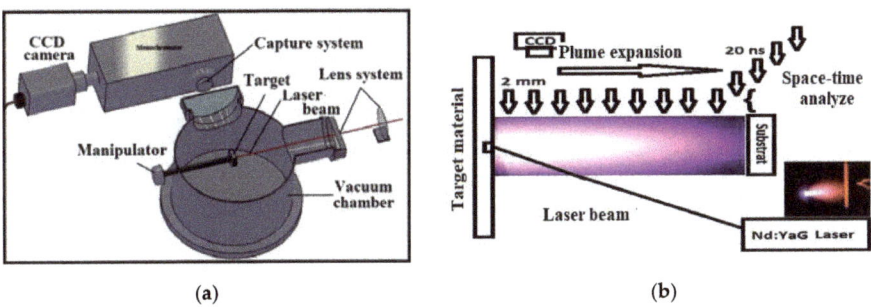

Figure 1. (a) Experimental deposition and analyze system and (b) time-distance analyze of the metallic plume.

Throughout the experiment, the substrate was maintained in a fixed position at room temperature. The target was fitted on a support of 45° that would direct the plasma of metallic ablated material towards the substrate. There were deposited a series of thin films from targets of Nitinol at two different distances between target and substrate (2 and 4 cm) using energies of laser of 40, 80, 100 mJ and two different deposition times (45 and 90 min). Given the substantial set of different deposition conditions attempted, a total of 30 samples were generated. In-situ plasma investigation was also implemented by means of both global and space- and time-resolved optical emission spectroscopy analyses. The global investigations focused on recording the overall emission of the plasma to analyze the overall dynamics of the ablated particle cloud while local investigations were performed by means of optical emission spectroscopy where the kinetics and thermal energy of individual species was investigated.

Usually there are two routes to achieve a shape memory crystallin thin films: one using a heated substrate and a second with post annealing heat treatment. Because the films deposited at high substrate temperatures (400–600 °C) are crystalline, they do not require any post-annealing treatments. However, initial tests have revealed that the heated substrate route was not resulting in high quality films, as the thin films became unstable at high temperatures and the growth of the thin film was accompanied by the formation of precipitates due to interfacial chemical reactions; also, these layers were more susceptible to cracking. Due to these reasons, in this case we chose to do a post deposition heat treatment of annealing by heating the layers to 800 °C in Ar atmosphere and cool them in air in order to crystalize the thin film structure.

The behavior of the shape memory materials that present transformations in a solid state was investigated in initial form and also of thin layer form through differential calorimetry on equipment DSC Maya 200 with a heating speed of 10 K/min in the atmosphere of Argon. The chemical analysis of the thin layers was performed through EDS analysis (detector Bruker: energy resolution up to 100 kcps count rate, use a self-calibrating P/B-ZAF analysis, minimum 90 nm diameter spot for analyze) and XRD (X'Pert Pro MRD, Scan-Continous, Start Angle: 20, End Angle: 120, Step size: 0.0131303, Time per step: 61.20, Scan speed: 0.05471, Number of steps: 7616, 45 KV, 40 mA, Anode X-ray Tube: Cu, 20:20–90) and the surface was analyzed through electron microscopy (SEM VegaTescan LMH II, work in a high vacuum of approximately 1×10^{-2} Pa, electron gun work 1–30 kV, 100,000×, 15.5 mm WD, SE detector and program VegaTC) and through atomic force microscopy (Nanosurf EasyScan 2 equipment, maximum Scan range: 110 µm, maximum Z-range: 22 µm, drive resolution Z: 0.34 nm, drive resolution XY 1.7 nm, XY-Linearity Mean Error: <0.6%, Z measurement noise level (RMS, Static Mode) 0.4 nm (max. 0.55 nm)).

3. Experimental Results

3.1. The Analysis of the Nitinol Target after the Ablation Process

The state of the surface of shape memory material that is used as a target in the process of depositing through laser ablation is presented in Figure 2 with the highlight of the traces left by the laser fascicle in Figure 2a and the thermic affected area in Figure 2b.

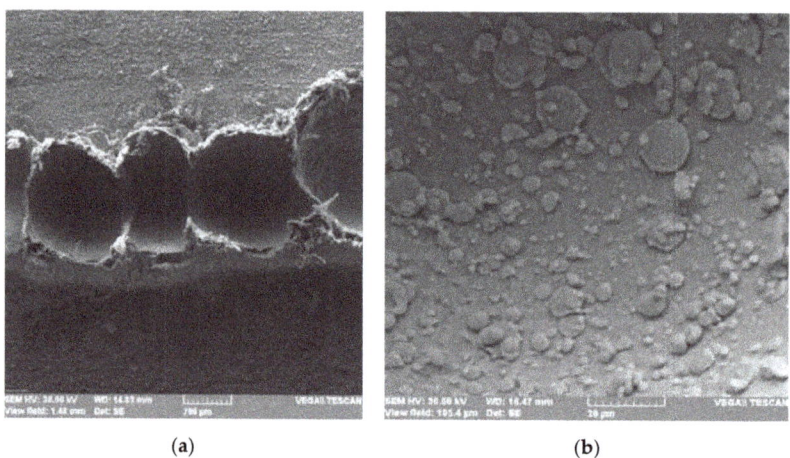

Figure 2. The state of the alloy surface NiTi after the laser exposure (**a**) perspective, (**b**) detail of area of ablated material.

We can observe an area of approximately 200 µm around the traces left by the laser, thermally affected by the laser action constituted from welded compounds of circa 1–3 µm. The action of laser in the contact area is characterized by a release of material on a surface of 314 µm² for a diameter of 20 µm as we can observe in Figure 2a. The surface affected by the laser beam was chemically analyzed on an area of 0.4 mm² by 10 t and the average of the values of identified chemical elements was of Ni: 47% wt., Ti: 41.5% wt. și O:11.5. The apparition of a high percentage of oxygen can be observed, a fact that signals a pronounced oxidation of the surface after the laser exposure with a decrease of the content of Ni that transformed at the surface in oxides. Another important aspect worth noting here is the fact that the heat affected area is large and there is a lot of redeposited material on the edges of the ablation crater. The presence of huge droplets on the targets is also a signature of the dominant effect induced by thermal ablation mechanisms, as opposed to the electrostatic ones. This is especially the case of explosive boiling, during which the ejection of large parts of materials can occur. This could translate into a high density of clusters or nanoparticles on the deposited films.

3.2. In-Situ Plasma Monitoring during Pulsed Laser Deposition of NiTi by Means of OES and ICCD Fast Camera Imaging

The central idea behind in-situ plasma monitoring during pulsed laser deposition of thin films is to potentially asses the quality and properties of the desired films by understanding the fundamentals of the deposition process. In order to obtain this, we performed in-depth characterization of irradiated target, resulted transient plasma and the thin films. One of the general accepted technique for in-situ plasma monitoring is optical emissions spectroscopy due to its non-perturbative nature. This approach is usually completed by ICCD fast camera imaging. Therefore, in order to understand the structure and the dynamics of the laser produced plasma, we employed a global to local approach. Global investigations involved ICCD fast camera imaging and the local ones the space and time resolved optical spectroscopy with the focus being on individual excited states seen in the ablation plasma.

The general dynamics of plasma (plume) resulted after the interaction between the laser fascicle and the target from Nitinol was analyzed through ICCD imaging of the laser produced plasma, with a gate width of 30 ns, for different moments of time. In Figure 3 there are presented some snapshots from the expansion of plasma in the precinct at different periods of time (from 25 to 1000 ns) after the impact between the laser fascicle and the target.

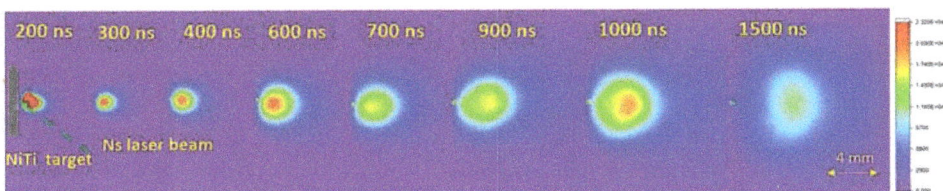

Figure 3. ICCD images of laser produced plasmas on NiTi shape memory target.

Starting with the first image where the plume was recorded at 200 ns after the interaction moment between the laser beam and the NiTi target (seen as $\Delta t = 0$), that the laser produced plasma has a quasi-spherical shape that increases in volume as it expands. At later moments, we noticed a split of the ablated cloud into three distinct structures (Figure 4). This separation is justified by the constituents of plume. The process was observed also in other studies [23,24] and can be attributed to a heterogeneity in the velocity of the ejected particles based on their nature and the ejection mechanisms involved.

The first structure, also known as the fast structure, is generated by electrostatic mechanisms (Coulomb explosion) and describes high energetic particles (mainly ions in various ionization states); the second structure is generated by thermal mechanism and usually describes a strongly atomized structure moving with a lower velocity, and the third structure, which is also of thermal nature and contains a heavier, more complex structure with a considerably lower expansion velocity [25]. In order to characterize the behavior of plume we estimated the velocities per each species by plotting the displacement of the maximum emission area in space and time. The values found were 1.6 km/s for the third structure, 2.8 km/s for the second one, and for the first one 5 km/s. The obtained value for the third structure is relatively high, meaning the ejected clusters have relatively high energy and there is a strong possibility of a high density of clusters on the deposited films. On the other hand the fast structure have a lower velocity in comparison to other reports from the literature, a sign that the forces driving the Coulomb explosion ejection mechanism are secondary to the thermal ones.

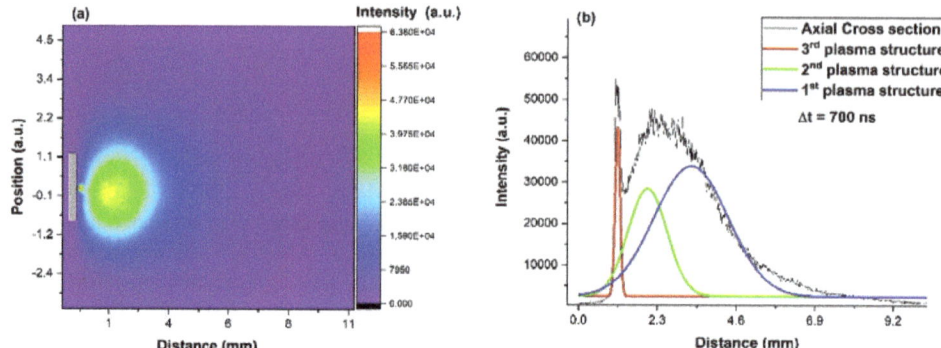

Figure 4. ICCD image of a laser produced plasma on NiTi recorded after 700 ns (**a**) and the cross section on the main expansion axis of the plume (**b**).

For the analysis of the contributions of the metallic species, which can be found in the plasma obtained from the NiTi target, a spectroscopic study was performed focusing on their evolution in space and time. The first global spectra of plasma for an area of 0.2 mm^2 from the plasma at a distance of 5 mm to the target surface was recorded in the 300–510 nm spectral range. In Figure 5 an example of spectra emitted for the NiTi sample is presented. Using appropriate databases [18] we identified all the emission lines observed empirically. We noticed both atomic and ionic Ti lines, and only atomic species of Ni. The abundance of Ti with respect to the Ni lines seen in the spectra can be explained by the lower evaporation heat and lower melting point of Ti coupled with an atomic radius almost three times higher than the Ni one. A higher atomic radius will translate into a higher collision cross-section of Ti, which in term could lead to the showcase of more excited states.

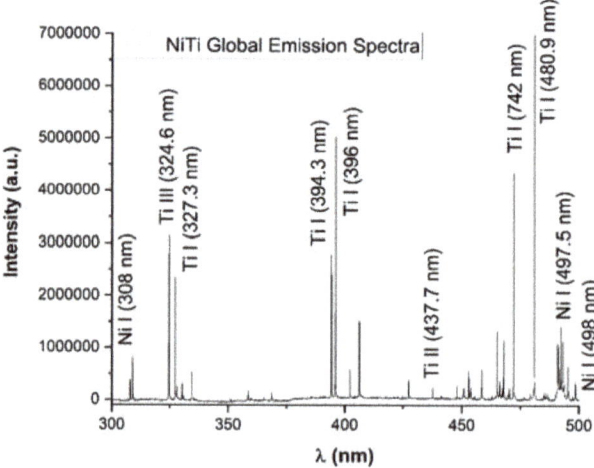

Figure 5. Global spectra for the TiNi laser produced plasma recorded with a gate width of 2 µs.

By performing space and time resolved optical emission spectroscopy and following the emission lines for the Ti and Ni atoms, we can determine according to the algorithm presented in [14] the expansion velocity of each species. We obtained for the Ni atom a velocity of 2.7 km/s, while for the Ti atoms a velocity of 5.7 km/s. We noticed that these velocities are close to the ones obtained by ICCD fast camera imaging. Thus, we can conclude that the splitting of the plume, in this case, is based on the nature of its components, as opposed to the mechanisms involved in the ablation process [14]. This is in good agreement with the high thermal effected area on the sample and with the mostly atomic emission lines observed, meaning that thermal mechanisms are dominant. With a dominant thermal ejection scenario, the composing particles of the ablated cloud will be separated on the basis of the individual thermal properties. This will result in a fast structure containing mainly Ti species, and a second structure mainly containing Ni species.

We further used the well-known Boltzmann plot method [14] to determine the individual excitation temperature of the two atomic species (Ni I and Ti I). For Ti I species we found an excitation temperature of 6500 K, while for Ni we found a relatively lower value of 7200 K, values determined at a distance of 0.5 mm from the target. This difference is in line with the differences in the thermal properties of the two components.

3.3. The Analysis of the Thin Layers Obtained through Laser Ablation from Targets with Shape Memory (NiTi)

The accurate control of the ratio Ni/Ti in the Ni-Ti films has an essential importance. The problems that may occur as seen through the plasma monitoring techniques is the differential evaporation of the two components from the target, coupled with the large scale non-uniformity of the PLD generated films, as well as structural changes to the target during the deposition process. In order to prevent these drawbacks, we implemented a combinatorial approach with the use of the NiTi target and simultaneously of a Ti sample to compensate the depleted Ti concentration from the films.

The resulted thin films were investigated by means of EDX technique. Selective spectra from the target and the thin films are presented in Figure 6. We observed a few peaks corresponding to the Ti and Ni species in the target and we have seen their corresponding peaks in the films with a reduced intensity. Due to this correspondence of the peaks, we can conclude that from a qualitative point of view, the stoichiometry of the ablated material is transferred to the substrate.

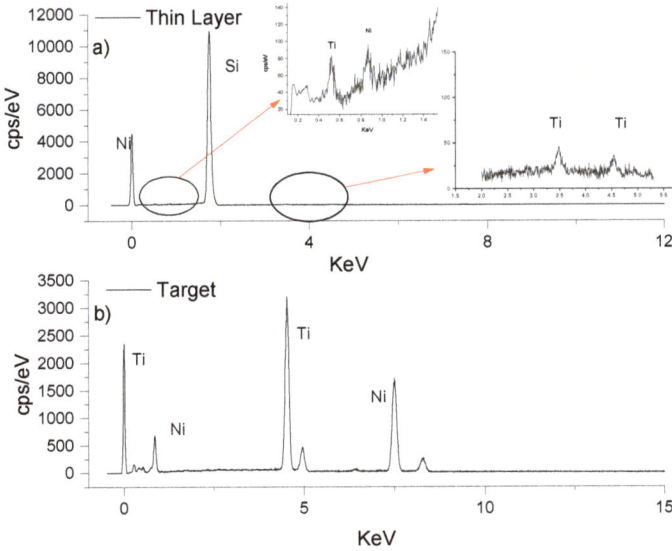

Figure 6. The spectra of energies realized on the EDS detector. (**a**) thin layer from NiTi and (**b**) target of Ni-Ti.

In order to determine the optimal conditions of depositing, but also to study the influence of the distance between the target and substrate on the properties of the films, we performed the study depicted in Figure 6. With this aim, areas of 0.4 mm² were analyzed from two samples obtained in different experimental conditions. The results obtained (in m/weight and atomic percentages) are presented in Table 1. The analyses were realized on a surface of approximately 1 mm² (Figure 7), which can contain various concentrations of clusters on their surfaces.

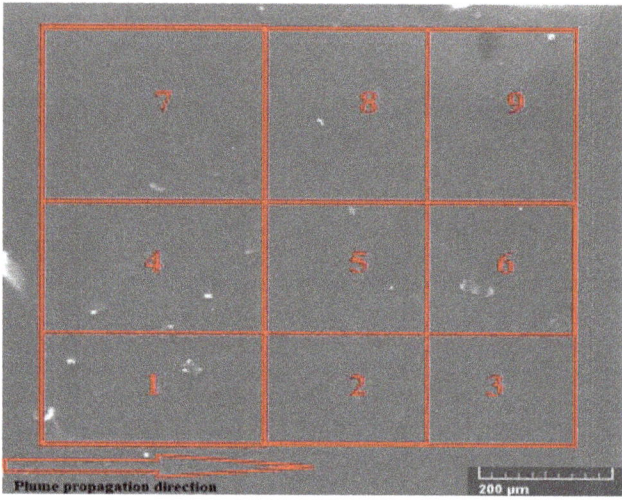

Figure 7. SEM image of the investigated thin surface and the zonal spread on which there were realized chemical analyses.

In Figure 7 it is specified the propagation direction of the material plasma from the target placed parallel with the sub layer of Si. The chemical analysis 1, 4, and 7 were selected on larger surfaces to determine the influence of the particle distribution within the plasma plume on the formation and development of the thin film. In the table is placed the chemical composition of target (Ni40Ti) used in the pulsed laser deposition process as reference value. We observed an area where we could achieve quasi-complete stoichiometric transfer in areas 4–6, situated in the center of the thin film, meaning that in the center of the plume keep the stoichiometry of the target while in the wider regions of the plasma some depletion of the Ti can occur. This result is in good agreement with those presented in [14] where it was shown that lighter elements are distribution mainly on the outer range of the plasma, as opposed to the heavier ones which are centered around the main expansion axis.

Table 1. The chemical composition of the thin film according to the distribution in Figure 7.

	Chemical Element		Target	Zone 1	Zone 2	Zone 3	Zone 4	Zone 5	Zone 6	Zone 7	Zone 8	Zone 9
Sample B	Ni (accuracy 0.2%)	wt.%	60.0	63.3	62.6	63.9	63.4	63.1	61.5	62.9	65.01	64.9
		at%	55.01	58.4	57.7	59.2	58.6	58.2	56.6	57.9	60.25	60.2
	Ti accuracy 0.15%)	wt.%	40.0	36.7	37.4	36.01	36.6	36.9	38.5	37.1	34.9	35.1
		at%	44.99	41.6	42.3	40.8	41.4	41.8	43.3	42.1	39.7	39.8
Sample A	Ni (accuracy 0.2%)	wt.%	60.0	61.3	60.9	58.7	60.1	58.74	58.1	58.2	59.6	57.3
		at%	55.01	56.4	55.9	53.6	55.1	53.7	53.1	53.2	54.6	52.3
	Ti (accuracy 0.15%)	wt.%	40.0	38.7	39.1	41.4	39.9	41.3	41.9	41.8	40.4	42.7
		at%	44.99	43.6	44.1	46.4	44.9	46.3	46.9	46.8	45.4	47.7

The standard variation of the Ni and Ti elements, from the base material (target), were determined: Ni: ±0.2 and Ti: ±0.1. In the case of TiNi thin film (sample a) positioned at a distance of 40 mm to the NiTi target, it was observed good stoichiometric transfer is achieved for almost all areas, most of them in the same experimental conditions in which sample B (20 mm) was also produced. The average atomic variation was in this case much lower than in the majority of the performed experiments, indicating a stoichiometric transfer according to the requirements of obtaining thin films with shape memory and with a percentage value higher than the data reported in the field of growing thin films through other methods such as thermic atomization.

In order to investigate the NiTi films formed on the Si base we used the XRD technique. Using this method, we could preserve equivalence between the base material (target with shape memory) and the material obtained on a substrate of Si. In Figure 8 a part of the XRD analyses performed on the obtained thin films is presented. Figure 8a highlights the characteristic peaks of a shape memory alloy NiTi in martensitic state at the room temperature. According to the database of the analysis equipment (ICC Database) [26], we established the main characteristic peaks of this material and the compound phases with their properties. In Figure 8b the spectrum obtained on a thin film is presented, for the NiTi alloy transported through laser ablation in amorphous state where we can find the majority of the thin layers obtained through laser ablation. Furthermore, we analyzed thin film through X ray diffraction, but after we applied a treatment of re-annealing of re crystallization, the result being presented in detail in Figure 8c.

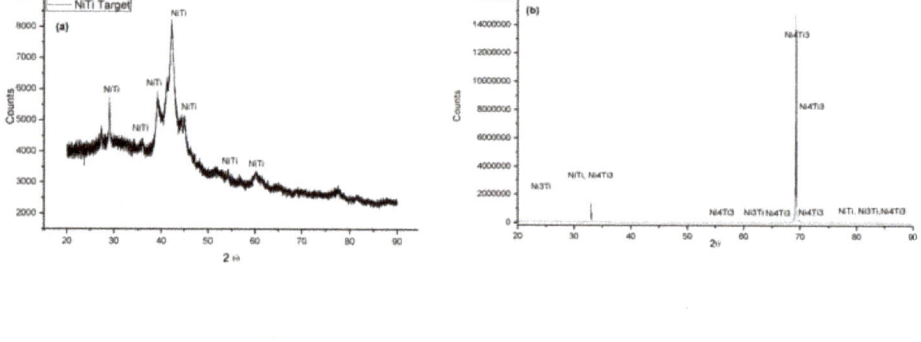

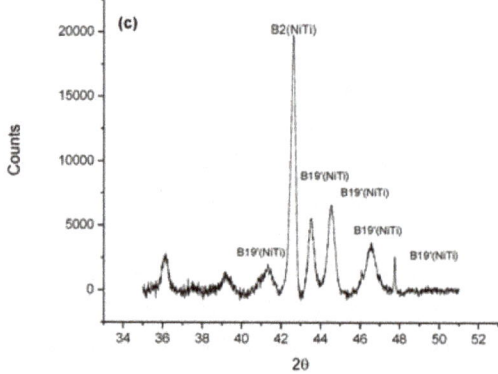

Figure 8. XRD analysis of (a) NiTi target, (b) TiNi thin film and (c) detail of specimen on heat treated TiNi thin film.

We observed numerous peaks specific to crystallized NiTi alloy. The size of the peaks was very low because the tests were performed in difficult conditions given by the reduced thickness of the layer (around 200 nm). The apparition of the peaks B2 and B19 characteristic to the austenitic and martensitic phase (in a very high percentage) were observed. After the annealing treatment, the material had a structure transformed to martensite. The formation of precipitates was not reported after the annealing treatment [27]. The appearance of untransformed phase: B_2, Figure 8c, was connected to the very fast formation of an Ni-rich layer near the oxide surface during the heat treatment.

In the case of shape memory films deposited under vacuum, the chemical composition of the new layers was influenced by the substrate temperature. Our results suggest that the chemical composition of a Nitinol layer can be precisely controlled without involving the substrate temperature using pulsed laser deposition but considering the distance between the target and substrate, and the orientation of the substrate contact surface with the plume.

3.3.1. Calorimetric Study

The removal of the samples from the Si layer was performed through debarking, and after the mass analyses the samples with the mass between 4 and 32 mg were investigated. Among these, only on the TiNi thin film crystallized through thermic treatment and with a net mass of 4 mg, an intern endotherm mass appeared around the temperature of 333 K (60 °C). The variation of the temperature flow in the case of this sample is presented in Figure 9.

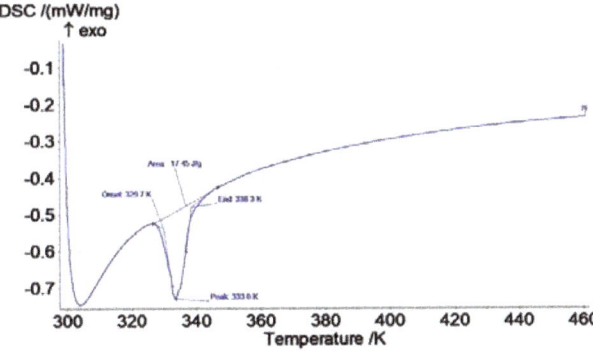

Figure 9. The variation of the heat flow depending on temperature of the exfoliated layer of NiTi.

All the other DSC analyses did not present transformations in the analysis area TC- 473 °C, but there is the possibility that will be taken into consideration in the future, for this domain to be too small and the transformation temperatures to be shifted even to negative temperatures in the case of NiTi alloys.

3.3.2. Surface Analyses

The condition of the surface of shape memory alloys is very important given that the relief state of the martensitic phase plays a major role in manifesting the shape memory effect [28]. We observed a pleated relief, characteristic for the martensitic variants, with an accurate arrangement of these and similar spatial dimensions with those resulted through SEM electronic microscopy. Also, a characteristic height of the profile with an average of 262 nm and a width of 2 μm of the martensitic plates is highlighted. The state of the thin film surface deposited through laser ablation from the target of Nitinol on the TiNi thin films (sample A) is presented in Figure 10 where we observe a typical relief to the variants of martensitic plates. The analyzed area does not contain large formations of material. The relief of the deposited thin layer is very low with an average value of 92.9 nm, Figure 10c a thermic treatment of partial recrystallization and annealing at 1073 K.

Using the specialized soft Vega TC of the scanning microscope, we followed morphologically the state of the surface of the material layer. In Figure 11 it is presented the state of the thin layer obtained through PLD from NiTi targets where we can notice the apparition of the characteristic drops of the depositing process through laser ablation. Depending on the depositing parameters and the ablated material, different morphologies of the thin layer surface presented. Only formations larger than 1 µm were analyzed, the rest being considered a part of the layer or too small to influence the characteristics of shape memory thin layer realized. All samples were investigated on the same material surface and in the same experimental conditions.

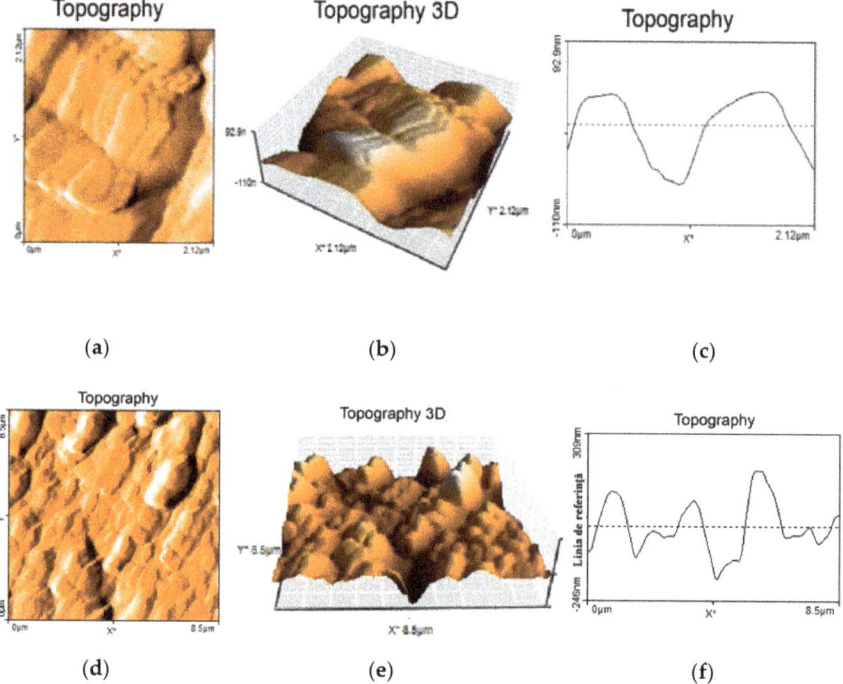

Figure 10. The AFM analysis of the thin layer surface on the sample A 14 after depositing (**a**) topography, (**b**) topography 3D and (**c**) linear topography. The AFM analysis of the thin layer surface on the sample B after depositing (**d**) topography, (**e**) topography 3D and (**f**) linear topography.

From the microstructural dimensional analysis (Figure 11) of the thin layers we observed that on the sample B were noticed 17 large particles (with an average of 2.62 µm) and the rest small particles under 1 µm (it is mentioned that the tests took place in all the cases on the surface of 115 × 115 µm). On the sample A there were observed 8 large particles with an average of 1.43 µm and the rest smaller than a micron. After the crystallization process normally, the grains grow isotopically until they press on each other or until they reach the layer surface or the substrate.

In the case of the layers obtained for a distance of 2 cm between target and substrate, the following values were obtained: at 40 mJ → 30 large particles with an average of 1.56 µm, at 80 mJ → approximately 50 large particles with an average of 2.39 µm and at the 100 mJ → appeared only 7 particles with an average of 2.02 µm.

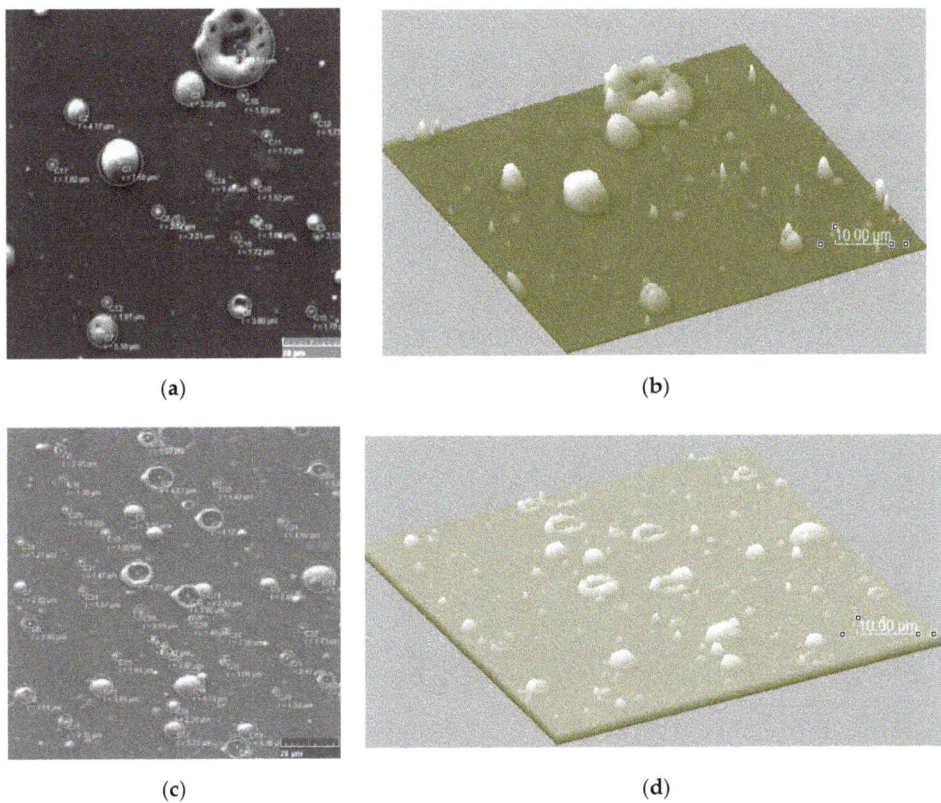

Figure 11. SEM microscopies for the sample A (**a**) 2D and (**b**) 3D for sample B (**c**) 2D and (**d**) 3D.

The influence of material drops on the properties of the submicronic layer deposited is given generally by differences of chemical composition that appear among these. In the case of shape memory alloys transformed in thin layers, it is essential to establish the behavior of the two materials that assemble the final layer (the layer and the material drops), the influence of the drops on the general properties of the thin layer, the increase or decrease of the number of micrometric droplets. For this reason, in the future we will pursue the investigation of these characteristics in order to increase the quality of thin layers with special properties.

4. The Dynamics of Transient Plasmas Generated by Laser Ablation of Memory Shape Alloy in a Fractal Paradigm

The laser-produced plasmas are non-differentiable (fractal) media induced by the collisions between the composing structural units (atoms, clusters, nanoparticles, etc.). The behavior of such a medium can be explicitly described in the Fractal Theory of Movement through dynamics on continuous and nondifferentiable curves (fractal curves [29,30]). In the following we will specify several types of nondifferentiable dynamics (either through hydrodynamic type regime at various scale resolutions, or through Schrodinger type regime at various scale resolutions) which will later be calibrated to the experimental results presented in the previous section.

If we further assimilate the ablation plasma with a complex system, the dynamics of its structural units can be described through the scale covariant derivative [30]:

$$\frac{\hat{d}}{dt} = \partial_t + \hat{V}^l \partial_l - i\lambda (dt)^{(\frac{2}{D_F})-1} \partial_l \partial^l \qquad (1)$$

where

$$\hat{V}^l = V^l - iU^l, \; i = \sqrt{-1} \qquad (2)$$

$$\partial_t = \frac{\partial}{\partial t}, \; \partial_l = \frac{\partial}{\partial X^l}, \; \partial_l \partial^l = \frac{\partial^2}{\partial X^l}\left(\frac{\partial}{\partial X^l}\right) \qquad (3)$$

In (1)–(3) X^l is the fractal spatial coordinate, t is the nonfractal temporal coordinate having the role of an affine parameter of the movement curves, $\hat{V}^l$ is the complex velocity, V^l is the real, differentiable, component of the velocity, independent of the scale resolution dt, U^l is the imaginary (nondifferentiable) component of velocity, dependent on the scale resolution, λ is the "diffusion coefficient" associated to the fractal-nonfractal transitions and D_F is the movement curve fractal dimension. For the previously defined fractal dimension we can choose any accepted definition: in a Kolmogorov sense, in a Haussdorff-Besikovici etc. [31]. However, once one definition is accepted it has to be constant throughout the study: $D_F < 2$ for correlative type processes, $D_F < 2$ for uncorrelated type processes etc.

The movement equation (geodesics equation) can be derived by implementing the scale covariance principle [29]. Then, through (1) and (2) it results:

$$\frac{\hat{d}\hat{V}^i}{dt} = \partial_t \hat{V}^i + \hat{V}^l \partial_l \hat{V}^i - i\lambda (dt)^{(\frac{2}{D_F})-1} \partial_l \partial^l \hat{V}^i = 0 \qquad (4)$$

This means that in any point of the movement trajectory the fractal acceleration, $\partial_t \hat{V}^i$, the fractal convection, $\hat{V}^l \partial_l \hat{V}^i$, and the fractal dissipation, $\partial_l \partial^l \hat{V}^i$ are reaching an equilibrium.

For the particular case of the irrotational movement:

$$\hat{V}^i = -2i\lambda (dt)^{(2/D_F)-1} \partial^i \ln \Psi \qquad (5)$$

where Ψ is the fractal state function and $\ln \Psi$ is the complex scalar potential of the complex velocity field, the movement Equation (4) takes the from of Schrödinger equation of fractal type [30]:

$$\lambda^2 (dt)^{(4/D_F)-2} \partial^l \partial_l \Psi + i\lambda (dt)^{(2/D_F)-1} \partial_t \Psi = 0 \qquad (6)$$

Moreover, by writing Ψ in an explicit manner:

$$\Psi = \sqrt{\rho} \exp(is) \qquad (7)$$

where $\sqrt{\rho}$ is an amplitude and s is a phase, the movement Equation (6) with the following substitutions:

$$\begin{aligned} V^l &= 2\lambda (dt)^{(\frac{2}{D_F})-1} \partial_l s \\ U^l &= \lambda (dt)^{(\frac{2}{D_F})-1} \partial_l \ln \rho \end{aligned} \qquad (8)$$

can be simplified by separating the dynamics of the ablation system on scale resolution. We obtain the equation system of fractal hydrodynamics:

$$\partial_t V^l + V^l \partial_l V^l = -\partial_l Q \qquad (9)$$

$$\partial_t \rho + \partial_l (\rho V^l) = 0 \qquad (10)$$

with Q the specific fractal potential:

$$Q = \frac{U_l U^l}{2} + \frac{\lambda(dt)^{(2/D_F)-1}}{2} \partial_l U^l = -2\lambda^2 (dt)^{(4/D_F)-2} \frac{\partial_l \partial^l \sqrt{\rho}}{\sqrt{\rho}} \qquad (11)$$

Equation (9) defined the specific momentum conservation law, while Equation (10) defines the state density conservation law. The specific fractal potential is a measure of the fractalization of the movement trajectory.

In the unidimensional case, the equation system of the fractal hydrodynamics:

$$\partial_t V + V \partial_x V = 2\lambda^2 (dt)^{(4/D_F)-2} \partial_x \left(\frac{\partial_x \partial_x \sqrt{\rho}}{\sqrt{\rho}} \right) \qquad (12)$$

$$\partial_t \rho + \partial_x (\rho V) = 0 \qquad (13)$$

with $V_x \equiv V$ and adequate initial and boundary conditions admits the analytical solution:

$$V = \frac{V_0 \alpha^2 + \mu^2 x t}{\alpha^2 + \mu^2 t^2} + i\lambda \frac{(x - V_0 t)}{\alpha^2 + \mu^2 t^2} \qquad (14)$$

$$\rho = \frac{(\pi)^{-1/2}}{[\alpha^2 + \mu^2 t^2]^{1/2}} \exp\left[-\frac{(x - V_0 t)}{\alpha^2 + \mu^2 t^2} \right] \qquad (15)$$

$$\mu = \frac{\lambda (dt)^{(\frac{2}{D_F}-1)}}{\alpha} \qquad (16)$$

in (14)–(16) V_0 is the initial velocity of a Gaussian packet described by $\rho(x,0) = \rho_0 \exp\left[-(x/\alpha)^2\right]$.

In such a context, using the second relation (8), the nondifferentiable velocity $U_x \equiv U$, has the expression:

$$U = \alpha \mu \frac{x - V_0 t}{\alpha^2 + \mu^2 t^2} \qquad (17)$$

From (17) and (14) it results the space-time homographic transformation:

$$x = \frac{U V_0 \alpha^2 + \alpha \mu V V_0 t}{\mu \alpha V + \mu^2 U t} \qquad (18)$$

Since at a differentiable resolution scale the following constriction is satisfied $\frac{V}{U} \gg \mu \alpha^{-1} t$, (18) takes the following form:

$$x \cong \frac{\alpha^2}{\lambda (dt)^{(\frac{2}{D_F}-1)}} \frac{U V_0}{V} + V_0 t \qquad (19)$$

which specifies a linear relationship of $x = x(t)$ type, having an unrestricted term of invers proportionality with the fractalization degree $\lambda(dt)^{(\frac{2}{D_F}-1)}$. In the following we will calibrate the theoretical model on the real dynamics of the laser produced plasma. Therefore, if we associate the fractalization degree with the nature of the atoms composing the target, through (19) the theoretical model can be corelated with the technique used to determine the expansion velocity from the ICCD fast camera imaging results.

Let us rewrite the homographic transformation under the following form:

$$x(t) = \frac{\alpha t + \beta}{\gamma t + \delta} \qquad (20)$$

Since the optical emission spectroscopy measurement implies a "spatial simultaneity", $dx = 0$, then this restriction can be expressed through differential form:

$$dt = \Omega^1 t^2 + \Omega^2 t + \Omega^3 \tag{21}$$

where $\Omega^1, \Omega^2, \Omega^3$ are the differential 1-forms of the SL(2R) algebra:

$$\Omega^1 = \frac{\alpha d\gamma - \gamma d\alpha}{\alpha\delta - \beta\gamma}, \quad \Omega^2 = \frac{\alpha d\delta - \delta d\alpha + \beta d\gamma - \gamma d\beta}{\alpha\delta - \beta\gamma}, \quad \Omega^3 = \frac{\beta d\gamma - \beta d\alpha}{\alpha\delta - \beta\gamma} \tag{22}$$

If there is a continuum parameter θ towards which the 1-forms are total differentials:

$$\Omega^1 = a^1 d\theta, \quad \Omega^2 = 2a^2 d\theta, \quad \Omega^3 = a^3 d\theta \tag{23}$$

with a^1, a^2, a^3 constants (21) can pe reduced to a Riccati type equation:

$$\frac{dt}{d\theta} = a^1 t^2 + 2a^2 t + a^3 \tag{24}$$

Let us know find the solution of (24) for a specific case:

$$A \frac{dt}{d\theta} = t^2 - 2Bt - AC \tag{25}$$

where we sued the following notations:

$$A = \frac{1}{a^1}, \quad B = -\frac{a^2}{a^1}, \quad AC = -\frac{a^3}{a^1} \tag{26}$$

with $a^1 > 0$. Admitting now that the roots of the following polynom:

$$P(t) = t^2 - 2Bt - AC \tag{27}$$

are given by the relation:

$$t_1 = B + iAk, \quad t_2 = B - iAk, \quad k = \frac{C}{A} - \left(\frac{B}{A}\right)^2 \tag{28}$$

The homographic substitution:

$$z = \frac{t - t_1}{t - t_2} \tag{29}$$

transforms (25) in:

$$\frac{dz}{d\theta} = 2ikz \tag{30}$$

The solution of this equation takes the following form:

$$z(t) = z(0) \exp(2ik\theta) \tag{31}$$

or even more:

$$t = \frac{t_1 + r \exp(2ik\theta) t_2}{1 + r \exp(2ik\theta)} \tag{32}$$

where r is a real constant specific to this particular solution. Using the relation (28), solution (32) in real terms becomes:

$$z = B + Ak \left[\frac{2r \sin(2k\theta)}{1 + r^2 + 2r\cos(2k\theta)} + i \frac{1 - r^2}{1 + r^2 + 2r\cos(2k\theta)} \right] \tag{33}$$

which will specify a self-modulation of the k characteristic through a Stoller type transformation [32]. In Figure 12a,b we represent the 3D and contour plot representation of $Re((z-B)/A$ as a function of r and $2k\theta$ for a scale resolution given by the maximum value of k. If we manage to calibrate this fractal representation onto the dynamics of laser produced plasmas, we have to identify the $Re((z-B)/A$ with the overall spectral emission of the plasma (in non-dimensional coordinates), with $2k\theta$ being the non-dimensional time and r being a non-dimensional specific length of the plasma (considered to the distance across the main expansion axis). We can see in Figure 12 that the for the overall emission of the plasma, a self-structuring phenomenon occurred. Each of the self-generated patterns characterizes a plasma structure which is generated by means of different ablation mechanism. Therefore, each structure will be characterized by a different fractalization degree. The connection between generation of self-structuring patterns and each observed plasma structure underlined the connection between the ablation mechanism and the fractalization degree was previously reported by our group [33].

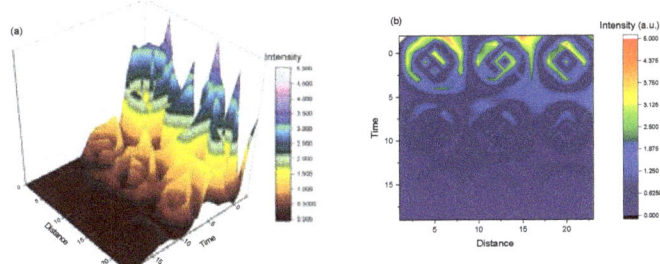

Figure 12. 3D (**a**) and contour plot (**b**) representation of $Re((z-B)/A$ assimilated with the global spectral emission of laser produced plasmas.

5. Conclusions

In order to obtain thin layers that can keep the properties of the bulk material, depositing technique through laser ablation was proposed. The NiTi targets oxidized in the impact area during the depositing process, impacting the particle removal mechanism during the ablation process. The plasma produced by laser ablation split during expansion in three structures. Each structure expands with different velocities correlated with the abundance of specific species presented in each structure. The kinetic properties of the structures were corelated with the individual kinetic and thermal properties of the Ti and Ni species. The chemical composition of the deposited layer depended of the distance of placing of the sub layer from the target and of the incidence angle between the material plasma and sub layer. The experimental results suggest that the chemical composition of a Nitinol layer can be precisely controlled without involving the substrate temperature using pulsed laser deposition considering the distance between the target and substrate, and the orientation of the substrate contact surface with the plume. The number of droplets on the deposited thin film decreased with the distance between the target and sub layer, and value of the laser fluence. Stoichiometric transfer was achieved for a series of samples.

A fractal model for describing the laser produced plasma dynamic, through various non-differential functionalities, was developed. Non-differentiable functionalities used for describing the laser produced plasmas dynamics through hydrodynamic type regime could be reduced to space-time homographic transformation. Transformations were correlated with the technique used to determine the global dynamics through ICCD fast camera imaging. Spatial simultaneity of the homographic type transformation implies through a special SL(2R) invariance dynamics defined through Riccati type equations. Such dynamics were correlated with optical emission spectroscopy measurements.

Author Contributions: Conceptualization, S.G.; N.C. and G.B.; methodology. G.B.; S.G.; N.C. and M.A.; investigation, G.B.; R.C.; N.C.; M.A. and S.G.; writing—original draft preparation, S.A.I.; R.C.; M.A.; writing—review and editing, S.A.I., V.-P.P.; M.A. and S.G.; visualization, S.A.I.; N.C.; R.C.; supervision, M.A.; S.G. and V.-P.P. All authors have read and agreed to the published version of the manuscript.

Funding: This work was supported by the Romanian Space Agency (ROSA) within Space Technology and Advanced Research (STAR) Program (Project no.: 114/7.11.2016) and the National Authority for Scientific Research and Innovation in the framework of Nucleus Program -16N/2019.

Conflicts of Interest: The authors declare no conflict of interest. The funders had no role in the design of the study; in the collection, analyses, or interpretation of data; in the writing of the manuscript, or in the decision to publish the results.

References

1. Sun, L.; Huang, W.M.; Ding, Z.; Zhao, Y.; Wang, C.C.; Purnawali, H.; Tang, C. Stimulus-responsive shape memory materials: A review. *Adv. Mater. Res. Switz.* **2012**, *33*, 577–640. [CrossRef]
2. Gomidzelovic, L.; Pozega, E.; Kostov, A.; Vukovic, N.; Krstic, V.; Zivkovic, D.; Balanovic, L. Thermodynamics and characterization of shape memory Cu–Al–Zn alloys. *Trans. Nonferrous Met. Soc. China* **2015**, *25*, 2630–2636. [CrossRef]
3. Bulai, G.; Trandafir, V.; Irimiciuc, S.A.; Ursu, L.; Focsa, C.; Gurlui, S. Influence of rare earth addition in cobalt ferrite thin films obtained by pulsed laser deposition. *Ceram. Int.* **2019**, *45*, 20165–20171. [CrossRef]
4. Balanovic, L.; Zivkovic, D.; Manasijevic, D.; Minic, D.; Cosovic, V.; Talijan, N. Calorimetric investigation of Al–Zn alloys using Oelsen method. *J. Therm. Anal. Calorim.* **2014**, *118*, 1287–1292. [CrossRef]
5. Mihaela, R.; Dascălu, G.; Stanciu, T.; Gurlui, S.; Stanciu, S.; Istrate, B.; Cimpoesu, N.; Cimpoesu, R. Preliminary Results of FeMnSi+Si(PLD) Alloy Degradation. *Key Eng. Mater.* **2015**, *638*, 117–122.
6. Ok Cha, J.; Hyun Nam, T.; Alghusun, M.; Sun, J. Composition and crystalline properties of TiNi thin films prepared by pulsed laser deposition under vacuum and in ambient Ar gas. *Nanoscale Res. Lett.* **2012**, *7*, 37. [CrossRef] [PubMed]
7. Ishida, A.; Martynov, V. Sputter-Deposited Shape-Memory Alloy Thin Films: Properties and Applications. *MRS Bull.* **2002**, *27*, 111–114. [CrossRef]
8. Miyazaki, S.; Fu, Y.Q.; Huang, W.M. *Thin Film Shape Memory Alloys: Fundamentals and Device Applications*; Cambridge University Press: New York, NY, USA, 2009.
9. Smausz, T.; Kecskeméti, G.; Kondász, B.; Papp, G.; Bengery, Z.; Kopniczkyd, J.; Hopp, B. Nanoparticle generation from nitinol target using pulsed laser ablation. *J. Laser Micro Nanoeng.* **2015**, *10*, 171–174. [CrossRef]
10. Suru, M.G.; Lohan, N.M.; Pricop, B.; Spiridon, I.P.; Mihalache, E.; Comaneci, R.I.; Bujoreanu, L.G. Structural effects of high-temperature plastic deformation process on martensite plate morphology in a Fe-Mn-Si-Cr SMA. *Int. J. Mater. Prod. Technol.* **2015**, *50*, 276–288. [CrossRef]
11. Cimpoeşu, N.; Stanciu, S.; Vizureanu, P.; Cimpoeşu, R.; Achiței, C.D.; Ioniță, I. Obtaining shape memory alloy thin layer using PLD technique. *J. Min. Metall. Sect. B Metall.* **2014**, *50*, 69–76. [CrossRef]
12. Bulai, G.; Gurlui, S.; Caltun, O.F.; Focsa, C. Pure and rare earth doped cobalt ferrite laser ablation: Space and time resolved optical emission spectroscopy. *Dig. J. Nanomater. Biostruct.* **2015**, *10*, 1043–1053.
13. Irimiciuc, S.; Bulai, G.; Agop, M.; Gurlui, S. Influence of laser-produced plasma parameters on the deposition process: In situ space- and time-resolved optical emission spectroscopy and fractal modeling approach. *Appl. Phys. A Mater. Sci. Process.* **2018**, *124*, 1–14. [CrossRef]
14. Irimiciuc, S.A.; Nica, P.E.; Agop, M.; Focsa, C. Target properties–plasma dynamics relationship in laser ablation of metals: Common trends for fs, ps and ns irradiation regimes. *Appl. Surf. Sci.* **2020**, *506*, 144926. [CrossRef]
15. Huang, W.M.; Ding, Z.; Wang, C.C.; Wei, J.; Zhao, Y.; Purnawali, H. Shape memory materials. *Mater. Today* **2010**, *13*, 54–61. [CrossRef]
16. Liu, C.; Qin, H.; Mather, P.T. Review of progress in shape-memory polymers. *J. Mater. Chem.* **2007**, *17*, 1543–1558. [CrossRef]
17. Available online: https://www.saesgetters.com/ (accessed on 9 October 2019).
18. Kramida, A.; Ralchenko, Y.; Reader, J. NIST ASD Team, NIST Atomic Spectra Database Lines Form, NIST at Spectra Database (Ver. 5.2). 2014. Available online: http://physics.nist.gov/asd (accessed on 9 October 2019).

19. Huang, W. On the selection of shape memory alloys for actuators. *Mater. Des.* **2002**, *23*, 11–19. [CrossRef]
20. Lagoudas, D.C. *Shape Memory Alloys: Modeling and Engineering Applications*; Springer: Boston, MA, USA, 2008.
21. Otsuka, K.; Ren, X. Physical metallurgy of Ti-Ni-based shape memory alloys. *Prog. Mater. Sci.* **2005**, *50*, 511–678. [CrossRef]
22. Sun, L.; Huang, W.M.; Cheah, J.Y. The temperature memory effect and the influence of thermo-mechanical cycling in shape memory alloys. *Smart Mater. Struct.* **2010**, *19*, 055005. [CrossRef]
23. Irimiciuc, S.; Boidin, R.; Bulai, G.; Gurlui, S.; Nemec, P.; Nazabal, V.; Focsa, C. Laser ablation of $(GeSe_2)_{100-x}(Sb_2Se_3)_x$ chalcogenide glasses: Influence of the target composition on the plasma plume dynamics. *Appl. Surf. Sci.* **2017**, *418*, 594–600. [CrossRef]
24. Irimiciuc, S.; Gurlui, S.; Agop, M. Particle distribution in transient plasmas generated by ns-laser ablation on ternary metallic alloys. *Appl. Phys. B* **2019**, *125*, 190. [CrossRef]
25. Canulescu, S.; Papadopoulou, E.L.; Anglos, D.; Lippert, T.; Schneider, C.W.; Wokaun, A. Mechanisms of the laser plume expansion during the ablation of $LiMn_2O_4$. *J. Appl. Phys.* **2009**, *105*, 063107. [CrossRef]
26. Istrate, B.; Mareci, D.; Munteanu, C.; Stanciu, S.; Luca, D.; Crimu, C.I.; Kamel, E. In vitro electrochemical properties of biodegradable ZrO_2-CaO coated MgCa alloy using atmospheric plasma spraying. *J. Opt. Adv. Mater.* **2015**, *17*, 1186–1192.
27. Wang, X.; Bellouard, Y.; Vlassak, J.J. Laser annealing of amorphous NiTi shape memory alloy thin films to locally induce shape memory properties. *Acta Mater.* **2005**, *53*, 4955–4961. [CrossRef]
28. Nandini, P.; Gagrani, A.; Singh Vipul, R.; Palani, A. Investigations on the Influence of Liquid-Assisted Laser Ablation of NiTi Rotating Target to Improve the Formation Efficiency of Spherical Alloyed NiTi Nanoparticles. *J. Mater. Eng. Perform.* **2017**, *26*, 4707–4717. [CrossRef]
29. Nottale, L. *Scale Relativity and Fractal Space-Time: An Approach to Unifying Relativity and Quantum Mechanics*; Imperial College Press: London, UK, 2011.
30. Merches, I.; Agop, M. *Differentiability and Fractality in Dynamics of Physical Systems*; World Scientific: Singapore, 2016.
31. Mandelbrot, B. *The Fractal Geometry of Nature*; WH Freeman Publisher: New York, NY, USA, 1993.
32. Stoler, D. Equivalence Classes of Minimum Uncertainty Packets. *Phys. Rev. D* **1970**, *1*, 3217. [CrossRef]
33. Irimiciuc, S.A.; Bulai, G.; Gurlui, S.; Agop, M. On the separation of particle flow during pulse laser deposition of heterogeneous materials—A multi-fractal approach. *Powder Technol.* **2018**, *339*, 273–280. [CrossRef]

© 2020 by the authors. Licensee MDPI, Basel, Switzerland. This article is an open access article distributed under the terms and conditions of the Creative Commons Attribution (CC BY) license (http://creativecommons.org/licenses/by/4.0/).

MDPI
St. Alban-Anlage 66
4052 Basel
Switzerland
Tel. +41 61 683 77 34
Fax +41 61 302 89 18
www.mdpi.com

Symmetry Editorial Office
E-mail: symmetry@mdpi.com
www.mdpi.com/journal/symmetry

www.ingramcontent.com/pod-product-compliance
Lightning Source LLC
LaVergne TN
LVHW070545100526
838202LV00012B/380